Advancement of Data Processing Methods for Artificial and Computing Intelligence

RIVER PUBLISHERS SERIES IN COMPUTING AND INFORMATION SCIENCE AND TECHNOLOGY

Series Editors

K.C. CHEN
National Taiwan University,
Taipei, Taiwan

University of South Florida,
USA

SANDEEP SHUKLA
Virginia Tech,
USA

Indian Institute of Technology Kanpur,
India

The "River Publishers Series in Computing and Information Science and Technology" covers research which ushers the 21st Century into an Internet and multimedia era. Networking suggests transportation of such multimedia contents among nodes in communication and/or computer networks, to facilitate the ultimate Internet.

Theory, technologies, protocols and standards, applications/services, practice and implementation of wired/wireless

The "River Publishers Series in Computing and Information Science and Technology" covers research which ushers the 21st Century into an Internet and multimedia era. Networking suggests transportation of such multimedia contents among nodes in communication and/or computer networks, to facilitate the ultimate Internet.

Theory, technologies, protocols and standards, applications/services, practice and implementation of wired/wireless networking are all within the scope of this series. Based on network and communication science, we further extend the scope for 21st Century life through the knowledge in machine learning, embedded systems, cognitive science, pattern recognition, quantum/biological/molecular computation and information processing, user behaviors and interface, and applications across healthcare and society.

Books published in the series include research monographs, edited volumes, handbooks and textbooks. The books provide professionals, researchers, educators, and advanced students in the field with an invaluable insight into the latest research and developments.

Topics included in the series are as follows:-

- Artificial Intelligence
- Cognitive Science and Brian Science
- Communication/Computer Networking Technologies and Applications
- Computation and Information Processing
- Computer Architectures
- Computer Networks
- Computer Science
- Embedded Systems
- Evolutionary Computation
- Information Modelling
- Information Theory
- Machine Intelligence
- Neural Computing and Machine Learning
- Parallel and Distributed Systems
- Programming Languages
- Reconfigurable Computing
- Research Informatics
- Soft Computing Techniques
- Software Development
- Software Engineering
- Software Maintenance

For a list of other books in this series, visit www.riverpublishers.com

Advancement of Data Processing Methods for Artificial and Computing Intelligence

Editors

Seema Rawat

Amity University in Tashkent, Uzbekistan

V. Ajantha Devi

AP3 Solutions, Chennai, India

Praveen Kumar

Astana IT University, Kazakhstan

NEW YORK AND LONDON

Published 2024 by River Publishers
River Publishers
Alsbjergvej 10, 9260 Gistrup, Denmark
www.riverpublishers.com

Distributed exclusively by Routledge
605 Third Avenue, New York, NY 10017, USA
4 Park Square, Milton Park, Abingdon, Oxon OX14 4RN

Advancement of Data Processing Methods for Artificial and Computing Intelligence / Seema Rawat, V. Ajantha Devi and Praveen Kumar.

Routledge is an imprint of the Taylor & Francis Group, an informa business

ISBN 978-87-7004-017-4 (hardback)
ISBN 978-87-7004-056-3 (paperback)
ISBN 978-10-0381-095-7 (online)
ISBN 978-1-032-63021-2 (ebook master)

While every effort is made to provide dependable information, the publisher, authors, and editors cannot be held responsible for any errors or omissions.

Contents

9   Method for Implementing Time-Control Functions in Real-time Operating Systems   **219**

*Wilver Auccahuasi, Oscar Linares, Karin Rojas, Edward Flores,
Nicanor Benítes, Aly Auccahuasi, Milner Liendo,
Julio Garcia-Rodriguez, Grisi Bernardo, Morayma Campos-Sobrino,
Alonso Junco-Quijandria and Ana Gonzales-Flores*

Preface

This book aims to showcase the applications of advancements in data processing methods for artificial intelligence (AI) in today's rapidly changing world, and how these advancements can benefit society through research, innovation, and development in this field. The book is relevant to a variety of data that contribute to the challenges of data science. With the rapid development and integration of technology in every aspect of life, the world has become a global village, and technology continues to shape the way we experience and live our lives. The book highlights the impact of this new-age technology, particularly in the areas of big data, engineering, and data science, and how it is ready to aid us in finding solutions to the challenges we face. The book aims to provide an in-depth understanding of the current state of AI and data processing technologies and their potential to drive future research and development in this field.

Section 1: Trends in Data Processing and Analytics

1. **Novel nonparametric method of multivariate data analysis and interpretation:**
   A nonparametric method for multivariate data analysis, interpretation, and classification based on the concept of tolerance ellipsoids is proposed. The method involves using multivariate tolerance regions in the form of special ellipsoids that cover a set of points. These ellipsoids have two unique features: 1) their significance level only depends on the number of sample values and 2) only one point is located on the surface of the ellipsoids. The chapter describes the algorithms for constructing these ellipsoids in high-dimensional spaces, the significance level of the ellipsoids, the computational complexity of the algorithm, and its practical usefulness. Additionally, the chapter investigates the statistical properties of the proposed method and introduces a new function for statistical depth of multivariate random values, a new method for multivariate ordering and statistical peeling, and a method for treating the uncertainty of the classification of points using these ellipsoids. The

proposed method is useful for analyzing and interpreting multivariate data, and classifying random multivariate samples, and may be applied in various fields such as medicine diagnostics, econometrics, image analysis, etc.

2. **Data analysis on automation of purging with IoT in FRAMO Cargo Pump:**

   A research study on a system used in oil and chemical tankers to transport cargo from ship to shore is analyzed in this chapter. The system, known as the FRAMO system, utilizes cargo pumps that are installed inside cargo tanks. One important aspect of maintaining the integrity of the system is performing purging procedures to confirm the integrity of the sealing arrangements on both the cargo and hydraulic systems. The proposed method in this paper uses three sensors – a density meter, pH meter, and a color sensor – to identify the liquid in the cofferdam space and distinguish between cargo liquid and hydraulic oil used in the system. The research aims to use physicochemical data to predict oil content and employs data mining techniques to classify the detected liquids with high accuracy.

3. **Big data analytics in the healthcare sector: potential strength and challenges:**

   The use of big data analytics in healthcare institutions and hospitals is discussed. With the rapid advancement in data acquisition and sensing technologies, healthcare institutions are now collecting large amounts of data about their patients, but this data is diverse in terms of its format, type, and the rate at which it needs to be analyzed. Conventional tools and methodologies are not sufficient to handle such vast volume of unstructured data; thus, sophisticated analytical tools and technologies are required, particularly big data analytics techniques, to interpret and manage the data related to healthcare. Big data analytics can be used in healthcare for better diagnosis, disease prevention, telemedicine, monitoring patients at home, and integrating medical imaging for wider diagnosis. The chapter aims to cover various problems and potential solutions that arise when big data analytics are utilized in medical institutions, and reviews big data analytical tools and methods in healthcare such as classification, clustering, artificial neural network, and fuzzy.

4. **Role of big data analytics in the cloud applications:**

   The combination of two technologies that are popular in today's IT industry: cloud computing and big data analytics. Big data refers to extremely large datasets that can be used to analyze underlying trends, patterns, and correlations that have business value. The process of analyzing patterns in big data using advanced analytical techniques is called big data analytics. Cloud computing, on the other hand, is the provision of computational services on demand over the internet. These services and the data stored are run on remote servers present in data centers across the world. The abstract explains how cloud computing and big data analytics can be combined to provide various services over the internet. Cloud computing enables the big data industry to achieve what it has not been able to in the past by providing a platform for real-time processing, interpretation, and delivery of data. Additionally, cloud computing's artificial intelligence and machine learning technologies can be used to convert unorganized data into a standard format. This can be very effective and economical for business enterprises as traditional on-premise data centers are harder to manage and need a lot of maintenance. The chapter concludes that by shifting their business model toward cloud, companies can achieve economies of scale with minimal investment and improve the overall customer experience.

5. **Big data analytics with artificial intelligence: a comprehensive investigation:**
    This chapter describes a survey of research on the use of artificial intelligence (AI) techniques in big data analytics. The rapid expansion of digital data, due to advancements in digital technologies such as the Internet of Things (IoT) and sensor networks, has resulted in the need for more effective techniques for analyzing large and complex datasets. Traditional relational databases and analytical methods are not able to handle such vast amounts of data; so new tools and analytical methods are needed to find patterns. AI techniques such as machine learning, knowledge-based, and decision-making algorithms can produce more accurate, quicker, and scalable results in big data analytics. The main objective of the survey is to examine AI-based Big data analytics research by using keywords to search relevant articles in databases such as ScienceDirect, IEEE Explore, ACM Digital Library, and SpringerLink. The selected articles are then grouped into categories of AI techniques and compared in terms of scalability, precision, efficiency, and privacy. The research period is between 2017 and 2022.

6. **Cloud computing environment for big data analytics:**
    The growing importance of effectively managing and extracting insights from large amounts of digital data, known as big data, is discussed in

this chapter. It mentions that the volume of data generated by various sources has increased significantly and that it can be difficult to effectively manage and evaluate. It also highlights that different methods and approaches of big data analytics can be utilized to extract valuable insights, which can improve decision-making. Additionally, the abstract describes how cloud computing, a powerful model for handling and storing large amounts of data, can be integrated with big data to provide scalability, agility, and on-demand data availability. The chapter also examines the definition and classification of big data, its tools and challenges, the definition, and types of cloud computing technology, and the architecture of cloud computing in big data technology, along with its advantages.

Section 2: Advance Implementation of Artificial Intelligence and Data Analytics

7. **Artificial intelligence-based data wrangling issues and data analytics process for various domains:**
    The role of data in day-to-day life, and how it is transformed from information to knowledge and wisdom, providing opportunities in digital technology is discussed in this chapter. The abstract highlights that companies that want to achieve success should depend on data as a fundamental resource for their operations and create data-driven business models. The use of artificial intelligence (AI) in data analytics is discussed as a way to handle large volumes of data by mimicking human brain functions and identifying data types, finding relationships between datasets, and recognizing knowledge. The abstract also mentions how AI can perform data exploration, speed up data preparation, automate processes, and create data models. Additionally, the use of natural language processing (NLP) in data analytics is discussed as a way to extract data, classify text based on keywords, and perform sentiment analysis. The abstract concludes by mentioning how AI is providing the best solution for data wrangling and analytics in various domains such as marketing, business, healthcare, banking, and finance

8. **Artificial intelligence and data science in various domains:**
    This chapter covers various aspects of artificial intelligence (AI). The chapter is divided into four parts. The first part provides an introduction to AI, its types, and its history. The second part discusses the ways to achieve AI along with its tools and working. The third part is about

the emergence of data science. The fourth and final part discusses the correlation between AI and its applications in various domains. The chapter includes examples of how AI can be implemented and used in various fields such as healthcare, finance, and transportation.

9. **Method for implementing time-control functions in real-time operating systems:**
   This chapter describes a proposed method for creating a real-time system based on the Linux kernel. The goal is to configure an operating system that allows for exclusive use of resources in order to execute applications with specific time and duration requirements. The method involves patching the Linux kernel and making various configurations. Example programs are also presented that demonstrate the use of different approaches for real-time programming. The method is intended to be scalable and can be applied to different processor architectures and devices for solving more complex control problems.

10. **Efficient blurred and de-blurred image classification using machine learning approach**
    A proposed approach for classifying and de-blurring images that have been deteriorated due to factors such as out-of-focus optics or variations in electrical imaging systems is explained in this chapter. The approach uses the discrete wavelet transform (DWT) to classify the blurry image and provides methods for de-blurring the image. The goal of this approach is to be able to distinguish blurred and unblurred images from input images. The proposed approach uses texture feature-based image classification using a neural network and machine learning approach. The texture features are extracted using gray-level co-occurrence matrix and an artificial neural network is used for classifying the images into different classes. The evaluation of the parameter analysis is also presented. The method is described to produce the best results.

11. **Method for characterization of brain activity using brain–computer interface devices:**
    A proposed method for using low-cost brain–computer interface (BCI) devices to record brain activity, which in the past would have required expensive medical equipment and had limited use is discussed in this chapter. The method uses solid-state electrodes placed on the forehead and an armor placed on the head, which is easy to use and has a wireless connection for transmitting signals to other devices. The abstract presents the protocol for using the equipment to obtain good records

and prepare the data for analysis, and also lists the different applications where brain activity can be recorded. The method is stated to be applicable, scalable, and adaptable to different needs.

12. **Streaming highway traffic alerts using Twitter API:**
    A proposed application for streaming analytics, which is a type of big data application, is covered in this chapter. The goal is to use real-time data analytics to forecast traffic by collecting traffic-related tweets from Twitter. The user would input a city name and the application would generate tweets about traffic in that city for the past seven days, alerting the user to avoid routes with blockages or accidents. The tweets are collected from the Twitter API using TweePy, and then classified as "traffic" or "non-traffic" using a model. The tweets also need to be pre-processed and lemmatized to improve the efficiency and accuracy of the model. The model is trained using a pre-classified dataset and then used to classify the tweets collected earlier. This chapter explains the data processing pipeline for streaming analytics using Twitter API and also summarizes the results and future directions for research.

13. **Harnessing the power of artificial intelligence and data science:**
    The impact of artificial intelligence (AI) and data science on various fields, specifically how it contributes to solving common problems, is discussed in this chapter. It is explained that AI and data science enable machines to exhibit human-like characteristics such as learning and problem-solving and that the healthcare and finance industries are examples of where it has a significant impact. The abstract also mentions that chatbots and virtual assistants are other examples of applications of AI and data science, where they are used for effective communication and completing tasks with simple voice commands. Additionally, it mentions that there are many other real-life applications of AI and data science such as e-mail spam filtering, recommendation systems, autocomplete, and face recognition. The abstract emphasizes that data science is the field of study of large amounts of data to discover previously unknown patterns and extract meaningful knowledge by combining domain expertise, using modern tools and techniques, and knowledge of mathematics and statistics.

14. **Determining the severity of diabetic retinopathy through neural network models:**
    This chapter describes a proposed approach for detecting diabetic retinopathy, a disease that affects the blood vessels of the light-sensitive

tissue in the eye and can lead to permanent loss of vision. The disease often has no symptoms in its early stages, making early detection crucial. The method involves using neural network models to classify the severity of the disease by analyzing retinal images. The models used in this study are VGG16, InceptionV3, ResNet50, and MobileNetV2. The study uses retinal images with varying properties such as contrast, intensity, and brightness to train the models and predict the stage of the disease. The results of the analysis show that the MobileNetV2 model had the highest training and testing accuracy. The abstract also presents accuracy graphs, loss graphs, confusion matrices, and classification reports for each model and for increasing the number of epochs, and the overall performance of all the models improves with an increase in the number of epochs.

15. **Method for muscle activity characterization using wearable devices:** The proposed solution for upper limb prosthesis that utilizes a wearable device for the simultaneous recording of muscle activity using electromyography (EMG) signals is covered in this chapter. The device has eight integrated acquisition channels and allows for wireless transmission of the signals to other devices. The abstract explains the acquisition protocol for recording arm muscles, and how each muscle's signal is separated. The proposed method can be used to help understand the behavior of certain muscles during certain activities, which can be used to improve the design of prostheses. The abstract also suggests using artificial intelligence techniques to recognize characteristic patterns in the recorded signals.

List of Contributors

Agarwal, Akshat, *Department of Computer Science and Engineering, Amity School of Engineering and Technology, Amity University Haryana, India*

Ajantha Devi, V., *AP3 Solutions, India*

Ankita, *Institute of Innovation in Technology and Management, India*

Auccahuasi, Aly, *Universidad de Ingeniería y Tecnología, Perú*

Auccahuasi, Wilver, *Universidad Científica del Sur, Perú*

Awasthi, A.K., *Department of Mathematics, School of Chemical Engineering and Physical Sciences, Lovely Professional University, India*

Awasthi, Monisha, *School of Computing Sciences, Uttaranchal University, India*

Ayushmaan, Das, *Sri Ramachandra Faculty of Higher Education and Research, India*

Benítez, Nicanor, *Universidad Nacional Mayor de San Marcos, Perú*

Bernardo, Grisi, *Cesar Vallejo University, Peru*

Bernardo, Madelaine, *Cesar Vallejo University, Peru*

Campos-Sobrino, Morayma, *Universidad Autónoma de Ica, Peru*

Choudhury, Tanupriya, *UPES, Bodholi Campus, India*

Flores, Edward, *Universidad Nacional Federico Villarreal, Perú*

Ganapathy, Jayanthi, *Sri Ramachandra Faculty of Higher Education and Research, India*

García-Rodríguez, Julio, *Universidad Privada Peruano Alemana, Peru*

Garov, Arun Kumar, *Department of Mathematics, School of Chemical Engineering and Physical Sciences, Lovely Professional University, India*

Ghosh, Kaushik, *UPES, Bodholi Campus, India*

Gonzales-Flores, Ana, *Universidad Autónoma de Ica, Peru*

Gowrishankar, R., *KIT-Kalaignarkarunanidhi Institute of Technology, India*

Gupta, Swati, *Department of Computer Science and Engineering, School of Engineering and Technology, K R Mangalam University, India*

Junco-Quijandria, Alonso, *Universidad Autónoma de Ica, Peru*

Klyushin, D.A., *Taras Shevchenko National University of Kyiv, Ukraine*

Liendo, Milner, *Escuela de Posgrado Newman, Perú*

Linares, Oscar, *Universidad Continental, Perú*

Madhu, *Institute of Innovation in Technology and Management, India*

Mahdi, Husain Falih, *Department of Computer and Software Engineering, University of Diyala Baquba, Iraq*

Malik, Sushma, *Institute of Innovation in Technology and Management, India*

Marque, Fausto Pedro Garcia, *University of Castilla-La Mancha, Spain*

Mohanakrishnan, Ramya, *Sri Ramachandra Faculty of Higher Education and Research, India*

Nanduri, Sriya, *Vellore Institute of Technology, India*

Pandimurugan, V., *SRMIST, India*

Prakash, P., *Vellore Institute of Technology, India*

Prasanna Kumar, R., *Indian Maritime University – Chennai Campus, India*

Pravakar, Devashree, *Vellore Institute of Technology, India*

Rajaram, V., *SRMIST, India*

Rana, Anamika, *Maharaja Surajmal Institute, India*

Rawat, Seema, *Amity University Tashkent, Uzbekistan*

Rodrigues, Paul, *Department of Computer Engineering, College of Computer Science, King Khalid University, Kingdom of Saudi Arabia (KSA)*

Rojas, Karin, *Universidad Tecnológica del Perú, Perú*

Sakthivel, V., *KADA, Konkuk University, Vellore Institute of Technology, India; Vellore Institute of Technology, India*

Saranya, G., *SRMIST, India*

Sharma, Minakshi, *Department of Mathematics, School of Chemical Engineering and Physical Sciences, Lovely Professional University, India*

Sharma, Sugandha, *UPES, Bodholi Campus, India*

Srividhya, S., *SRMIST, Kattankulthur, India*

Tymoshenko, A.A., *Taras Shevchenko National University of Kyiv, Ukraine*

Udayakumar, E., *KIT-Kalaignarkarunanidhi Institute of Technology, India*

Vetrivelan, P., *PSG Institute of Technology and Applied Research, India*

Vijarania, Meenu, *Department of Computer Science and Engineering, School of Engineering and Technology, K R Mangalam University, India*

Yadav, Aarti, *Department of Computer Science and Engineering, School of Engineering and Technology, K R Mangalam University, India*

List of Figures

List of Tables

List of Abbreviations

ACF	Autocorrelation function
AES	Advanced Encryption Standard
AI	Artificial intelligence
ANN	Algorithm of neural network
AR	Autoregressive
ARIMA	Autoregressive Integrated Moving Average
ATCS	Adaptive Traffic Control System
AWS	Amazon Web Services
BCI	Brain–computer interface
BDA	Big Data Analytics
BDaaS	Big data as a service
BDBN	Bilinear deep belief network
CAD	Computer-aided Design
CDH	Cloudera distributed Hadoop
CI	Computing intelligence
CIBIL	Credit Information Bureau (India) Limited
CLAHE	Contrast limited adaptive histogram equalization
CMAP	Concept Map
CNN	Convolutional neural network
CNT	Cognitive tool
CPU	Central processing unit
CRIF	Central Reference Information Facility
CSP	Cloud service providers
CSV	Comma-separated values
CTA	Classification tree analysis
CUDA	Compute Unified Device Architecture
DAIC	Defence Artificial Intelligence Council
DBaaS	Database as a Service
DBMS	Database management systems
DBN	Deep belief network
DL	Deep learning
DM	Diabetes mellitus

DNN	Deep neural network
DP	Deep learning
DR	Diabetic retinopathy
DRF	Discrete residual flow
DSR	Design science research
DWT	Discrete wavelet transform
ECG	Electrocardiogram
EEG	Electroencephalogram
EHB	E-Health and Bioengineering
EHR	Electronic health records
EM	Expectation maximization
EMG	Electromyography
ER	Emergency room
ETL (or ELT)	Extraction, transformation, and loading
FC	Fully connected
GAN	Generative adversarial network
GANN	Genetic algorithm neural network
GPU	Graphics processing unit
GRNN	General regression neural network
HDFS	Hadoop distributed file system
HMH	Hardin Memorial Health
HSTL	High-speed transistor logic
IaaS	Infrastructure as a service
ICISC	International Conference on Inventive Systems and Control
ICISS	International Conference on Intelligent Sustainable Systems
ILSVRC	ImageNet large scale visual recognition challenge
IMU	Inertial measurement unit
IoMT	Internet of Medical Things
IoT	Internet of Things
IRB	Institutional Review Board
IrMAs	Intraretinal microvascular abnormalities
ISRO	Indian space research organization
IT	Information technology
ITMS	Intelligent traffic management system
KNIME	Konstanz Information Miner
KNN	k-Nearest neighbor
LHS	Learning healthcare system
LRN	Local response normalization
LSTM	Long short-term memory

LVDCI	Low voltage digital control impedance
MA	Miniature aneurysms
MAP	Maximum a posteriori
MATLAB	Matrix laboratory
MC	Modulation classification
MEP	Motor evoked potential
MI	Motor images
ML	Machine learning
MLE	Maximum likelihood estimation
MPSO	Modified Particle Swarm Optimization
MX	Apache MXNet
MYO	A gesture control armband developed by Thalmic Labs
NER	Name entity recognition
NLP	Natural language processing
NN	Neural network
NPDR	Non-proliferative diabetic retinopathy
OCT	Optical coherence tomography
OS	Operating system
PaaS	Platform as a service
PACF	Partial autocorrelation function
PART	Partial tree
PDR	Proliferative diabetic retinopathy
PM	Pattern mining
POS	Part of speech
PSF	Point spread function
PSU	Public sector undertakings
RDBMS	Relational Database Management System
RDX	Remote discriminator network
ReLU	Rectified linear unit
RF	Random forests
RNN	Recurrent neural network
ROC	Receiver operating characteristic
SaaS	Software as a service
SCP	Slow cortical potentials
sEMG	Surface electromyography
SNA	Social network analysis
SNR	Signal-to-noise ratio
SQL	Structured query language
SSL	Secure sockets layer
SVM	Support vector machines

TDBN	Temporal deep belief networks
TMS	Transcranial magnetic stimulation
TNRDC	Tamil Nadu Road Development Corporation
VHF	Very high frequency
WNPRP	Wagon Next Point Routing Protocol
WPA	Wavelet packets

Introduction to Advancement of Data Processing Methods for Artificial and Computing Intelligence

V. Ajantha Devi

AP3 Solutions, India

Email: ap3solutionsresearch@gmail.com

Abstract

Artificial intelligence (AI) and computing intelligence (CI) have become increasingly important in recent years, with many businesses and organizations looking to leverage these technologies to improve their operations and gain a competitive edge. One of the key components of AI and CI is data processing, which involves the collection, cleaning, and analysis of data in order to train and improve AI and CI systems. There have been several key advancements in data processing methods for AI and CI in recent years [1]. Another important advancement has been the development of deep learning, which involves the use of neural networks with multiple layers to analyze and process data. This has led to the creation of even more powerful AI and CI systems that are able to process large amounts of data and make complex decisions.

Big data analytics is another area that has seen significant advancements. With the increasing amount of data being generated, traditional data processing methods are no longer sufficient. Big data analytics involves the use of advanced techniques and technologies to process, manage, and analyze large and complex datasets. This has led to the development of more accurate and efficient AI and CI systems [2]. The advancement of cloud computing has also had a significant impact on data processing for AI and CI. With the ability to store, process, and analyze large amounts of data in the cloud, organizations can more easily access, manage, and analyze large datasets and use them to train and improve their AI and CI systems.

1 Introduction

The field of artificial intelligence (AI) and computing intelligence (CI) has made tremendous strides in recent years, thanks in part to advancements in data processing methods. These methods involve the collection, cleaning, and analysis of data in order to train and improve AI and CI systems. These advancements have led to the creation of more sophisticated and accurate AI and CI systems that can process large amounts of data and make complex decisions. As the world becomes increasingly digital, the ability to effectively process, analyze, and make sense of vast amounts of data has become vital for businesses and organizations of all sizes. Artificial intelligence (AI) and computing intelligence (CI) are at the forefront of this revolution, enabling machines to learn, improve, and make decisions with minimal human intervention [3]. The success of these technologies relies heavily on the ability to process large amounts of data, and advancements in data processing methods have enabled AI and CI to reach new heights of performance and capabilities, shaping the future of these technologies [4].

1.1 Advancements in machine learning algorithms

The efficacy of machine learning algorithms has improved, and their use has grown across many industries as a result of recent major advancements. The use of deep learning methods like convolutional neural networks and recurrent neural networks has increased, which is one noteworthy development. With the help of these techniques, advances in voice and image identification as well as natural language processing have been made. Reinforcement learning has also made strides, and it has been used to teach agents how to carry out a variety of tasks, including playing challenging games and operating machines [5]. Progress has also been made in the area of generative models, which can produce fresh data that is comparable to an input. The development of systems that can carry out duties that were previously believed to be the sole responsibility of people as a result of these developments in machine learning algorithms has the potential to revolutionize many sectors.

1.2 Advancements in deep learning

Deep learning, a subset of machine learning, has seen significant advancements in recent years. One of the major breakthroughs has been in the field of convolutional neural networks (CNNs), which have led to improved performance in image and video recognition tasks. Another area of advancement has been in recurrent neural networks (RNNs), which have been used for natural language processing tasks such as language translation and text-to-speech synthesis [6].

Emergence of architectures such as Trans-former, BERT, GPT-2, and GPT-3 [7–9] is another field of development. These systems were able to produce state-of-the-art outcomes on natural language understanding tasks like question–answering, named object identification, and sentiment analysis. Overall, deep learning advancements have led to the development of systems that can perform tasks that were previously thought to be the exclusive domain of humans and have the potential to revolutionize many industries such as computer vision, natural language processing, and speech recognition [10, 11].

1.3 Advancements in big data analytics

Big data analytics is an ever-evolving field, and recent advancements have led to significant improvements in the ways in which data can be analyzed and understood. One major advancement is the increasing use of machine learning algorithms, which are able to identify patterns and insights in data that would be difficult or impossible for humans to detect [12]. Additionally, there has been a growing emphasis on real-time processing, allowing for near-instantaneous analysis of streaming data. Cloud computing has also become more prevalent in the field, providing scalable and cost-effective storage and processing power for large datasets. Another important development is the use of graph databases and graph analytics, which helps in understanding the relationship between the data. Overall, these advancements [13] have led to more efficient, accurate, and actionable insights from big data, with wide-ranging applications in industries such as finance, healthcare, and retail.

1.4 Advancements in cloud computing

Cloud computing has undergone significant advancements in recent years, leading to increased flexibility, scalability, and cost-effectiveness for businesses and organizations [14]. One major advancement is the development of multi-cloud and hybrid cloud strategies, which allow for a combination of public and private cloud resources to be used in a way that best suits the needs of the organization. Additionally, there has been a growing emphasis on security and compliance, with many cloud providers now offering advanced security features such as encryption and identity management. Another important development is the emergence of serverless computing, which allows for the execution of code without the need for provisioning or managing servers [15]. This leads to cost savings and increased scalability. The use of containers and container orchestration platforms such as Kubernetes has also become more prevalent, allowing for greater flexibility and portability in cloud-based applications. Overall [16], these advancements in cloud computing have led to increased adoption and reliance on cloud services, with wide-ranging applications in industries such as finance, healthcare, and retail.

2 Impact of Advancements on AI and CI Systems

The advancements in artificial intelligence (AI) and cognitive systems (CI) have had a significant impact on various industries and aspects of our daily lives. One major impact is the automation of repetitive and mundane tasks, leading to increased efficiency and productivity. Additionally, the use of AI and CI in decision-making has led to more accurate and informed decisions. With the advancements in natural language processing (NLP), CI systems have become better at understanding and responding to human language, leading to more human-like interactions. Another important impact is the ability of AI and CI systems to analyze and make sense of large amounts of data, leading to new insights and discoveries. AI and CI systems have become more prevalent in industries such as healthcare, finance, and retail, and are expected to play an increasingly important role in the future. However, it is also important to consider the ethical and social implications of the increased use of AI and CI systems and to ensure that they are developed and used in a responsible and inclusive manner.

In recent years, there have been significant advancements in the fields of machine learning (ML), deep learning (DL), big data analytics, and cloud computing that have greatly improved the capabilities of artificial intelligence (AI) and cognitive systems (CI). Together, these technologies have enabled the creation of more intelligent, efficient, and powerful AI and CI systems that can process and analyze vast amounts of data, learn from it, and make decisions in real time [17]. The advancements in ML algorithms and techniques have led to more accurate and efficient systems. Deep learning, a type of ML [18], utilizes neural networks with multiple layers to extract features and patterns from data.

Big data analytics is the process of collecting, storing, and analyzing large amounts of data to uncover hidden patterns, correlations, and insights. The advancements in big data analytics have led to more efficient and effective ways of processing and analyzing large amounts of data, which has greatly improved the performance of AI and CI systems. Cloud computing is the delivery of computing services over the internet, including storage, processing power, and applications [19]. The advancements in cloud computing have led to more cost-effective and scalable ways of storing and processing data, which has greatly improved the performance of AI and CI systems [20].

Together, these technologies have enabled the creation of more powerful and sophisticated AI and CI systems that can process and analyze vast amounts of data, learn from it, and make decisions in real time. This has led to wide-ranging applications in various industries such as healthcare, finance, and retail. For example, AI-powered systems are being used to improve the accuracy of medical diagnoses, and DL is being used to create more realistic and engaging virtual assistants. However, it is also important to consider the ethical and social implications of the increased use of AI and CI systems, and to ensure that they are developed and used in a responsible and inclusive manner. With the fast pace of development in these technologies, it is important to monitor and evaluate the impact they have on society and take necessary steps to mitigate any negative effects.

3 Organization of the Book

Data processing methods for artificial and computing intelligence have undergone significant advancements in recent years, driven by the explosion of data and the need for more sophisticated analysis techniques. These advancements have enabled organizations to leverage data in new and innovative ways, leading to improved decision-making, enhanced customer experiences, and increased efficiency.

Section 1: Trends in Data Processing and Analytics

This section will cover the latest trends in data processing and analytics, including big data, cloud computing, machine learning, and natural language processing. It will explore how these technologies are being used to transform industries and drive innovation. Additionally, it will highlight some of the key challenges that organizations face when working with large volumes of data and how they can overcome these obstacles.

Section 2: Advance Implementation of Artificial Intelligence and Data Analytics

This section will delve deeper into the advanced implementation of artificial intelligence (AI) and data analytics. It will discuss the different types of AI, such as deep learning and reinforcement learning, and their applications in various industries. Furthermore, it will cover advanced data analytics techniques, including predictive analytics, prescriptive analytics, and real-time analytics. It will also explore some of the ethical considerations and potential risks associated with these technologies.

Overall, this topic is essential for individuals and organizations looking to stay ahead of the curve in the rapidly evolving field of data processing and artificial intelligence. By understanding the latest trends and techniques, they can leverage data to drive innovation, gain a competitive edge, and deliver superior customer experiences.

In conclusion, the evolution of machine learning, deep learning, big data analytics, and cloud computing has greatly improved the capabilities of AI and CI systems. These technologies have led to more powerful, efficient, and intelligent systems that can process and analyze vast amounts of data, learn from it, and make decisions in real time. However, as these technologies continue to advance, it is important to consider their ethical and social implications and ensure that they are developed and used in a responsible and inclusive manner. In conclusion, ML, DL, BDA, and CC have all played an important role in improving AI and CI systems. Each of these technologies has enabled AI and CI systems to become more accurate and reliable, allowing them to make better decisions and predictions. As technology continues to evolve, it is likely that these technologies will continue to be used to improve AI and CI systems.

References

[1] Jordan, M. I., & Mitchell, T. M. (2015). Machine learning: Trends, perspectives, and prospects. Science, 349(6245), 255–260.

[2] Chen, M., Mao, S., & Liu, Y. (2014). Big data: A survey. Mobile Networks and Applications, 19(2), 171–209.

[3] Chen, T., & Xie, L. (2020). Advances in data processing methods for artificial intelligence and computing intelligence. In Advances in Artificial Intelligence, Software and Systems Engineering (pp. 105–118). Springer.

[4] Alom, M. Z., Rahman, M. M., Nasrin, M. S., Taha, T. M., & Asari, V. K. (2019). Advances in deep learning for big data analytics. Computational Intelligence, 35(3), 414–436.

[5] Silver, D., Huang, A., Maddison, C. J., Guez, A., Sifre, L., Van Den Driessche, G., ... & Dieleman, S. (2016). Mastering the game of Go with deep neural networks and tree search. Nature, 529(7587), 484–489.

[6] Lecun, Y., Bengio, Y., & Hinton, G. (2015). Deep learning. Nature, 521(7553), 436–444. doi: 10.1038/nature14539

[7] Devlin, J., Chang, M. W., Lee, K., & Toutanova, K. (2018). BERT: pre-training of deep bidirectional transformers for language understanding. arXiv preprint arXiv:1810.04805.

[8] Radford, A., Narasimhan, K., Salimans, T., & Sutskever, I. (2018). Improving language understanding by generative pre-training. OpenAI Blog, 1(8), 9.

[9] Brown, T. B., Mann, B., Ryder, N., Subbiah, M., Kaplan, J., Dhariwal, P., Neelakantan, A., Shyam, P., Sastry, G., Askell, A., Agarwal, S., Herbert-Voss, A., Krueger, G., Henighan, T., Child, R., Ramesh, A., Ziegler, D. M., Wu, J., … Amodei, D. (2020). Language models are few-shot learners. Advances in Neural Information Processing Systems, 33.

[10] Hochreiter, S., & Schmidhuber, J. (1997). Long short-term memory. Neural computation, 9(8), 1735–1780. doi: 10.1162/neco.1997.9.8.1735

[11] Vaswani, A., Shazeer, N., Parmar, N., Uszkoreit, J., Jones, L., Gomez, A. N., Kaiser, Ł., & Polosukhin, I. (2017). Attention is all you need. Advances in Neural Information Processing Systems, 5998–6008.

[12] Laney, D. (2012). "3D Data Management: Controlling Data Volume, Velocity, and Variety," Gartner Research.

[13] Liu, B., Li, H., & Zhu, X. (2015). "Big data: applications and opportunities in the healthcare industry." Journal of Healthcare Engineering, 6(2), 157–178.

[14] Armbrust, M., Fox, A., Griffith, R., Joseph, A. D., Katz, R., Konwinski, A., … & Zaharia, M. (2010). A view of cloud computing. Communications of the ACM, 53(4), 50–58.

[15] Boniface, M., Matthews, B., & Clayman, S. (2014). Cloud computing: principles, systems and applications. John Wiley & Sons.

[16] Li, J., Li, J., Li, K., & Li, Y. (2017). Security and privacy in cloud computing: A survey. International Journal of Distributed Sensor Networks, 13(3), 1550147717704543.

[17] Marr, B. (2015). Big data: Using smart big data, analytics and metrics to make better decisions and improve performance. John Wiley & Sons.

[18] Shalev-Shwartz, S., & Ben-David, S. (2014). Understanding machine learning: From theory to algorithms. Cambridge university press.

[19] Roberts, D. (2018). Serverless computing: Overview and recommendations for application development. Journal of Object Technology, 17(2), 2–13.

[20] Leitner, P., Cito, J., Gall, H. C., & Avritzer, A. (2018). Container orchestration for edge computing: Industry practices and research challenges. Proceedings of the 1st International Workshop on Edge Systems, Analytics and Networking, 1–6.

SECTION 1

Trends in Data Processing and Analytics

1

Novel Nonparametric Method of Multivariate Data Analysis and Interpretation

D.A. Klyushin and A.A. Tymoshenko

Taras Shevchenko National University of Kyiv, Ukraine

Email: dmytroklyushin@knu.ua

Abstract

The chapter describes a nonparametric method of multivariate data analysis, interpretation, and classification based on tolerance ellipsoids. We propose to use a multivariate tolerance region using special ellipsoids covering a set of points. These ellipsoids have two special features: 1) their significance level depends only on the number of sample values, and 2) only one point lies on their surface. We describe the algorithms of construction of the ellipsoids in high-dimension space. We demonstrate the significance level of these ellipsoids, the computational complexity of the algorithm, and its practical usefulness. Further, we investigate the statistical properties of the proposed method and introduce a novel function of statistical depth of multivariate random values, a novel method of multivariate ordering and statistical peeling, and a method of treating the uncertainty of the classification of points using such ellipsoids. The fact that only one point lies on the surface of such an ellipsoid allows for constructing concentric ellipsoids and estimating quantile levels, detecting outliers (anomalies), and unambiguous ordering of the points with respect to their statistical depth. In addition, this method allows raising the accuracy of classification and interpretation of points in high-dimensional space. A deeper point in the set has a higher rank. Comparing the rank of the point in every set, we can definitely interpret the point. The proposed method is useful for analyzing and interpreting multivariate data and classifying multivariate random samples. It may be applied in various domains, such as medical diagnostics, econometrical analysis, image analysis, etc.

1.1 Introduction

Numerous applications associated with data processing often require analysis of their statistical depth, particularly the determination of the most probable value (median) and the least probable values (outliers). The chapter proposes a new statistical depth function based on tolerance ellipsoids. This function is characterized by uniqueness, the accuracy of probability estimation, and relatively low computational complexity. This chapter explores the statistical properties of this function and demonstrates its effectiveness in analyzing the statistical depth of random multivariate data. Also, the concept of statistical depth is very close to the concept of prediction sets and the ordering of multivariate data.

Classification of the methods of multivariate ordering was proposed by Barnett [1]. These methods were classified as marginal, reduced, partial, and conditional. Marginal methods of ranking use separate components. Reduced methods of ranking estimate some distance from every sample to a central point. Partial ranking divides multidimensional samples into groups where samples are considered as equivalent. Finally, conditional ranking orders multidimensional samples using selected components, and a rule of ranking other elements depends on the selected component.

Different concepts of statistical depth were proposed by Tukey [2] (halfspace depth), Oja [3] (Oja depth), Liu [4] (simplicial depth), Koshevoy and Mosler [5] (zonoid depth), Rousseeuw and Hubert [6] (regression depth), Cascos [7] (convex hull depth), Lange and Mozharovsky [8] (zonoid depth), Lyashko et al. [9] (elliptical peeling), etc.

1.2 Statistical Depth

First, the concept of statistical depth was proposed by J. Tukey in 1975 [2]. An informative survey of multivariate depth functions is provided by Mosler and Mozharovskyi [10].

Consider a multivariate probability distribution P in R^d. A monotonic function $D(x,P)$ that depends of x and P and ranks random points obeying P such that it attends a maximum at the center point (median) is called a statistical depth function. The value $D(x,P)$ for x and P is called the depth of x. The deepest point is called a median. Depending on a definition of a statistical depth function, a median may be a geometrical center, centroid, etc.)

Desirable properties of a statistical depth function were proposed by Zuo and Serfling [4].

A1. Affinity invariance. The statistical depth function must be independent of affine transformations of data.

A2. Maximum in the center. The statistical depth function must attend the maximum value at the median.

A3. Monotonicity. The depth of points from considering distribution monotonically decreases relative to an imaginary line passing through the distribution center – the deepest point of distribution.

A4. Depth at infinity. If $\|x-x_{median}\| \to \infty$, then $D(x,P) \to 0$.

1.3 Prediction Sets

The main idea of a prediction set is the following. Consider independent identically distributed multivariate data $X_1, X_2, \ldots, X_n$, where $X_i \in R^d$. A prediction set constructed by these data is a region $\Omega(X_1, X_2,\ldots, X_n) \subset R^d$ such that $P(X_{n+1} \in \Omega) \geq 1 - \alpha$, where $0 < \alpha < 1$ and X_{n+1} is drawn from the same general population as $X_1, X_2, \ldots, X_n$ (in other words, $X_1, X_2, \ldots, X_n, X_{n+1}$ are independent identically distributed samples).

Prediction sets, or tolerance sets, were studied by Wilks [11], Wald [12], Tukey [13], Fraser and Guttman [14], Guttman [15], Aichison and Dunsmore [16], Bairamov and Petunin [17], Shafer and Vovk [18], Li & Liu [19], Lei et al. [20], Ndiaye and Takeuchi [21], and Lyashko et al. [9].

In this chapter, we describe and investigate the ideas of elliptical peeling based on Petunin ellipsoids.

1.4 Petunin Ellipsoids and Their Statistical Properties

The algorithm proposed by Yu. I. Petunin [22] constructs an ellipsoid covering a set of multivariate points. The original aim of this ellipsoid was to approximate a minimal volume ellipsoid covering a given set of random points. Indeed, the Petunin ellipsoid good approximates the minimal volume ellipsoid covering a given set of random points, but its main advantage is the fact that it provides a known probability that an arbitrary point from a given distribution belongs to the Petunin ellipsoid, and this probability is equal to $\frac{n}{n+1}$, where n is a number of points.

The simplest case of the Petunin ellipsoid that provided clear visualization is a Petunin ellipse (i.e., $d = 2$). Hereinafter, M_n is a set of random points (x_k, y_k) in R^2.

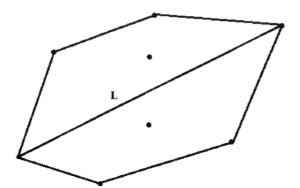

Figure 1.1   Two most distant points.

1.4.1 Case $d = 2$

Find a diameter of M_n. Let (x_k, y_k) and (x_1, y_1) be the ends of the diameter of M_n. It is well known that they are vertices of a convex hull of M_n. Construct a line L passing thought (x_k, y_k) and (x_1, y_1) (Figure 1.1).

Find the vertices of the convex hull (x_r, y_r) and (x_q, y_q) that are most distant from L on both sides. Construct lines L_1 and L_2 that pass through (x_r, y_r) and (x_q, y_q) and parallel to L. Construct lines L_3 and L_4 that pass through (x_k, y_k) and (x_1, y_1) and orthogonal to L (Figure 1.2). Lines L_1, L_2, L_3, and L_4 form a rectangular Π with sides a and b (let for definiteness $a \leq b$).

Rotate and translate the plane such that the left bottom vertex of the rectangle is an origin of a new coordinate system with axes Ox' and Oy', and points (x_1, y_1), (x_2, y_2), ..., are mapped to (x'_1, y'_1), (x'_2, y'_2), ..., (x'_n, y'_n) (Figure 1.3).

Transform abscises of points (x'_1, y'_1), (x'_2, y'_2), ..., (x'_n, y'_n) shrinking the rectangular with the coefficient $\alpha = a/b$. As a result, we have points $(\alpha x'_1, \alpha y'_1)$, $(\alpha x'_2, \alpha y'_2)$, ..., $(\alpha x'_n, \alpha y'_n)$ in a square S.

Find the center (x'_0, y'_0) of and compute distances $r_1, r_2, ..., r_n$ from the center to every point of the set. Find $R = \max (r_1, r_2, ..., r_n)$. Construct a circle with the center (x'_0, y'_0) and radius R (Figure 1.4).

Perform the inverse operation of stretching circles along the axe Ox' with the coefficient $\beta = 1/\alpha$ and inverse rotation and translation. As a result, we construct a tolerance ellipse. We call it a Petunin ellipse (Figure 1.5).

The computational complexity of the algorithm is determined by the computational complexity of constructing a convex hull of the points. Its estimate is $O(nlgn)$.

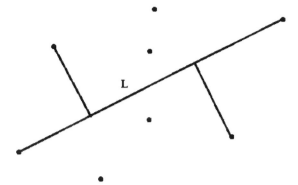

Figure 1.2 Two most distant points from *L*.

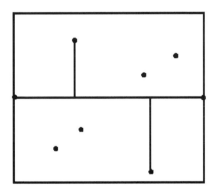

Figure 1.3 Translation, rotation, and scaling.

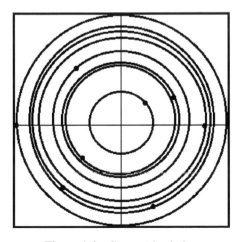

Figure 1.4 Concentric circles.

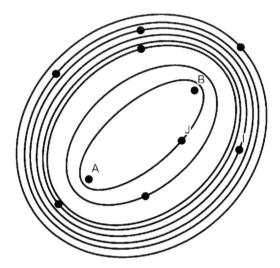

Figure 1.5   Concentric ellipses.

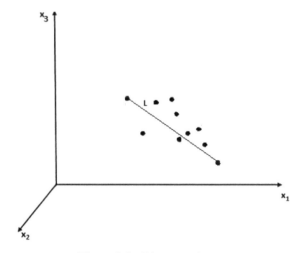

Figure 1.6   Diameter of a set.

1.4.2 Case *d* > 2

Find the diameter of M_n. Let X_k and X_i are vertices that are ends of the diameters. Construct a line L passing through X_k and X_i (Figure 1.6).

Rotate and translate the coordinate system so that the diameter of the convex hull is aligned along to Ox_1'. Project points to a plane that is orthogonal to Ox_1'. Perform convex hull construction, rotation, and translation in this subspace (Figure 1.7).

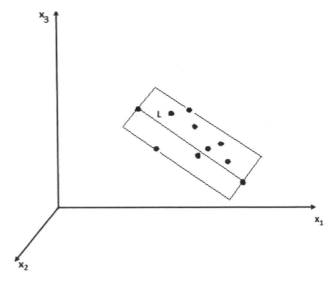

Figure 1.7 Projection.

Here, we obtain a rectangular covering all the projects of the points in the d-dimensional space. Therefore, we reduce a multidimensional problem to a one-dimensional one. Now, we may repeat the algorithm for the plane described above.

Repeat these steps while the dimension of a projection subspace does not equal to 1. Then, construct the least rectangular parallelepiped aligned along coordinate axes and covering points $X_1, X_2, ..., X_n$ (Figure 1.8).

Shrink the space transforming the parallelepiped to a hypercube (Figure 1.9). Find the center X_0 of this hypercube and compute distances $r_1, r_2, ..., r_n$ from it to every point. Find $R = \max(r_1, r_2, ..., r_n)$. Construct a hyperball with the center X_0 and the radius R. Performing the inverse operations of stretching, rotating, and translating, we obtain the desired ellipsoid in the initial space (Figure 1.10).

1.4.3 Statistical properties of Petunin`s ellipsoids

The Hill's assumption [23] states that if sample values $x_1, x_2, ..., x_n$ are exchangeable identically distributed random values obeying absolutely continuous distribution function, then,

$$P\left(x_{n+1} \in \left(x_{(i)}, x_{(j)}\right)\right) = \frac{j-1}{n+1},$$

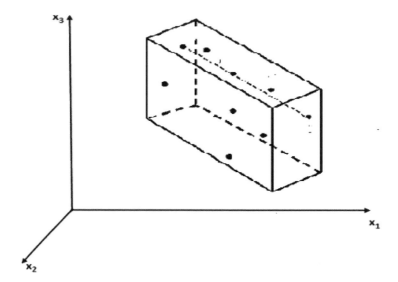

Figure 1.8 Parallelepiped.

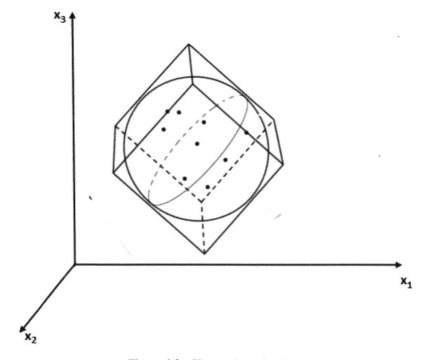

Figure 1.9 Hypercube and sphere.

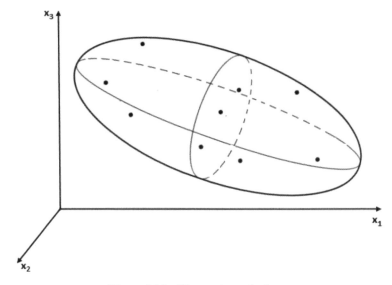

Figure 1.10 Hypercube and sphere.

where x_{n+1} obeys the same distribution as $x_1, x_2, \ldots, x_n$, and $x_{(i)}, x_{(j)}$ are the ith and jth order statistics.

The following theorems hold [8].

Theorem 1.1. If $x_1, x_2, \ldots, x_n, x_{n+1}$ are exchangeable random values obeying absolutely continuous joint distribution without ties then

$$P\left(x_i \geq x_k\right) = \frac{1}{n+1} \quad \forall k = 1,2,\ldots,i-1,i+1,\ldots,n+1.$$

Theorem 1.2. If $x_1, x_2, \ldots, x_n, x_{n+1}$ are exchangeable random values obeying absolutely continuous joint distribution function without ties, then

$$P\left\{x_{n+1} \in \left(x_{(j)}, x_{(j+1)}\right]\right\} = \frac{1}{n+1} \quad \text{and} \quad P\left\{x_{n+1} \in \left(x_{(i)}, x_{(j)}\right]\right\} = \frac{j-i}{n+1} \forall \, 0 \leq$$

$i < j \leq n+1.$

Theorem 1.3. If $X_1, X_2, \ldots, X_n$ are exchangeable multivariate random points in R^d, E_n is the Petunin ellipsoid covering $X_1, X_2, \ldots, X_n$, and X_{n+1} obeying the same distribution as $X_1, X_2, \ldots, X_n$, then $P\{X_{n+1} \in E_n\} = \dfrac{n-1}{n+1}.$

If $n > 19$, then the significance level of the Petunin ellipsoid is less than 0,05.

Note, that 1) only one point always lays at the surface of the Petunin ellipsoid; 2) the Petunin ellipsoid covers n random points with probability $\frac{n}{n+1}$; 3) to guarantee the significance level not greater than 0.05 a set of random points may contain not less 20 random samples.

Using the Petunin tolerance ellipsoids we can construct an elliptical statistical depth function interpolating the depth of every point in a set laying at the concentric ellipsoids. It is easy to see that the proposed elliptical statistical depth satisfies conditions A1–A4.

In addition, let us consider how the Petunin ellipsoids correspond to the concept of depth-ordered regions proposed by Cascos [7], where depth-ordered regions (central) are some analog of prediction sets:

$$D_\alpha(P) = \left\{x \in R^d : D(x,P) \ge \alpha\right\},$$

where $D(x,P)$ is a statistical depth of the multivariate point x taken from the distribution P. Cascos [7] introduced the desired features of deep-ordered regions.

1. *Affine equivariance*: $\forall x \in R^d\ D_\alpha(Ax + b) = AD_\alpha(x) + b$, where $A \in R^{d \times d}$ is a random nonsingular matrix and $b \in R^d$.

2. *Nesting*: if $\alpha \ge \beta$, than $D_\alpha(P) \subset D_\beta(P)$.

3. *Monotonicity*: if $x \le y$ by components, than $D_\alpha(P) + a \subset D_\beta(P) + a$, where and $a \in R^d$ is a random positive vector.

4. *Compactness*: $D_\alpha(P)$ is the compact area.

5. Convexity: $D_\alpha(P)$ is the convex area.

It is easy to see that the depth-ordered Petunin ellipsoid satisfies conditions 1–5.

1.5 Petunin Ellipsoids and Their Statistical Properties

In this research, 1000 points were used for creating Petunin ellipsoid (basic points), and another 1000 points generated with the same distribution were tested. Each distribution – exponential, gamma, Laplace, normal, and uniform had two different parameter combinations for more diverse testing. The test was performed 100 times for each combination to calculate average results and statistical properties. All resulting points were scaled into squares with side 100 in order to make resulting areas smaller, and a comparison of areas of resulting ellipses (generated by each point) was made. Results of this

Table 1.1 Statistical parameters of coverage level for uniformly distributed points.

Parameter for number of points inside divided by all points amount	Test 1 (equal sides)	Test 2 (rectangular)
Mean	0.99586	0.998
Mean deviation	0.00266	0.00135
Mode	0.996	0.99748
Median	0.997	0.999
Minimal	0.983	0.992
Maximal	1	1
Points outside (average)	5	2

testing are represented below – tables for statistical values and ellipse area comparison for each type of distribution.

1.5.1 Uniform distribution

As we see in the uniform case, Petunin ellipsoid covered most of the tested points for both tests. Even for the worst achieved result, only 1.7% of tested points were outside. All the tested points belong to 0...100 square area, so no extra scaling was applied here.

Figures 1.11 and 1.12 describe the area of ellipses, created according to basic points – from inner ellipse to ellipse covering all basic points and most tested points.

Removing points near corners of the area results in faster ellipse area reduction, but after ellipse is inside the square area, each next inner point creates a slightly smaller ellipse than the previous, so area reduction slows down.

1.5.2 Normal Distribution

As we see in the normal case, Petunin ellipsoid is really effective for normal distribution because the group of testing points looks like an ellipse itself. The worst achieved result is 1.1% of tested points outside – a really low percentage. Here scaling of generated and tested points into square 100 by 100 was used to make sure all points were inside.

Figures 1.13 and 1.14 describe the area of ellipses created according to basic points – from inner ellipse to ellipse covering all basic points and most tested points.

Here, the number of points far from the center of the generated set is small, so each point seriously influences the ellipse area. At the same time,

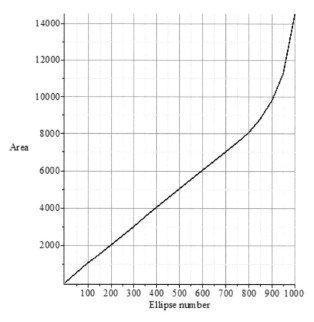

Figure 1.11   Area of Petunin ellipses covering 1000 uniformly distributed points sorted by radius – equal sides.

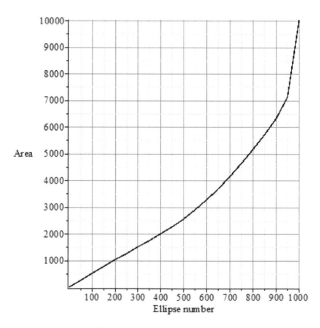

Figure 1.12   Area of Petunin ellipses covering 1000 uniformly distributed points sorted by radius – rectangular.

Table 1.2 Statistical parameters for normally distributed points

Parameter for number of points inside divided by all points amount	Test 1 (equal sides)	Test 2 (horizontal part longer)
Mean	0.99746	0.9979
Mean deviation	0.00174	0.00163
Mode	0.9965	0.99648
Median	0.998	0.999
Minimal	0.989	0.989
Maximal	1	1
Points outside (average)	3	3

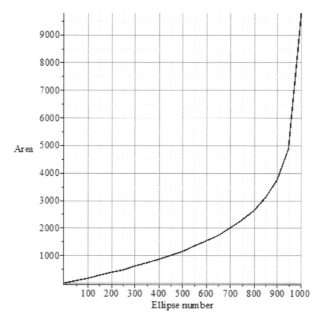

Figure 1.13 Area of Petunin ellipses covering 1000 normally distributed points sorted by radius – equal sides.

the density of points nearer to the central part is high – so ellipse areas are alike, and areas change slowly.

1.5.3 Laplace distribution

As for the Laplace distribution case, Petunin ellipsoid is still effective. Form of the testing point set reminds a rhombus with high density in the middle and small points far from the central part. The worst achieved result is 1.3% of tested points outside – so ellipses cover most of the points. Here, we used scaling of generated and tested points into square 100 by 100.

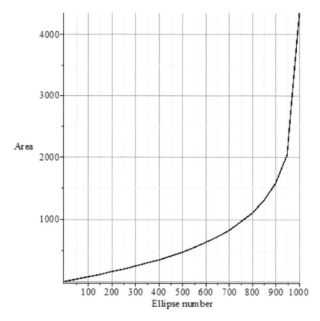

Figure 1.14   Area of Petunin ellipses covering 1000 normally distributed points sorted by radius – horizontal part longer.

Table 1.3   Statistical parameters of Laplace distributed points.

Parameter for number of points inside divided by all points amount	Test 1	Test 2
Mean	0.9968	0.9965
Mean deviation	0.0021	0.0019
Mode	0.998	0.995
Median	0.998	0.997
Minimal	0.987	0.988
Maximal	1	1
Points outside (average)	4	4

Figures 1.15 and 1.16 describe the area of ellipses, created according to basic points – from inner ellipse to ellipse covering all basic points and most tested points.

According to these figures, points far from the center of the generated set have the most serious influence on ellipse area – their density is low, and each adds large extra area. At the same time, for most other points, the density of points is high – so ellipse areas are alike and areas change slowly.

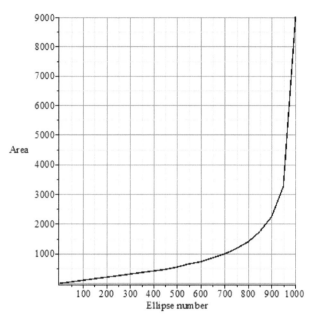

Figure 1.15 Area of Petunin ellipses covering 1000 Laplace distributed points sorted by radius – test 1.

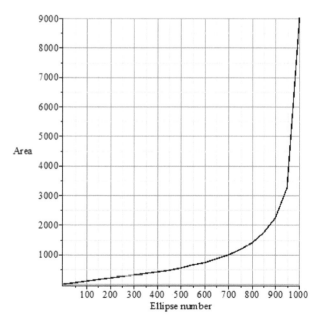

Figure 1.16 Area of Petunin ellipses covering 1000 Laplace distributed points sorted by radius – test 2.

Table 1.4   Statistical parameters of gamma-distributed points.

Parameter for number of points inside divided by all points amount	Test 1	Test 2
Mean	0.99775	0.9965
Mean deviation	0.00168	0.0019
Mode	0.9984	0.995
Median	0.998	0.997
Minimal	0.988	0.988
Maximal	1	1
Points outside (average)	3	4

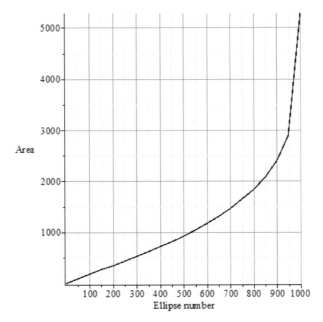

Figure 1.17   Area of Petunin ellipses covering 1000 gamma-distributed points sorted by radius – test 1.

1.5.4 Gamma distribution

For the gamma-distribution case, Petunin ellipsoid also demonstrated high accuracy. Form of the testing point set looks like a cloud with a low amount of points around and a high concentration near its central part. The worst achieved result is 1.2% of tested points outside – also a really low percentage. Scaling of generated and tested points into square 100 by 100 was used here too.

Figures 1.17 and 1.18 describe an area of ellipses, created according to basic points – from inner ellipse to ellipse covering all basic points and most tested points.

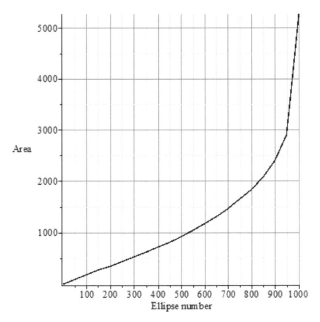

Figure 1.18 Area of Petunin ellipses covering 1000 gamma-distributed points sorted by radius – test 2.

Table 1.5 Statistical parameters of exponentially distributed points.

Parameter for number of points inside divided by all points amount	Test 1	Test 2
Mean	0.99824	0.99818
Mean deviation	0.00133	0.0015
Mode	0.99797	0.99797
Median	0.999	0.999
Minimal	0.993	0.992
Maximal	1	1
Points outside (average)	2	2

Here, area reduction is slower than the previous time, and the impact of points distant from the central part is high only for the first 50 points, then areas become nearer to each other. This distribution looks similar to normal for area comparison.

1.5.5 Exponential distribution

For the exponential distribution case, Petunin ellipsoid covered the majority of points, and the testing point set reminds a corner with a high density of points near zero, medium density around, and several points far from it. The

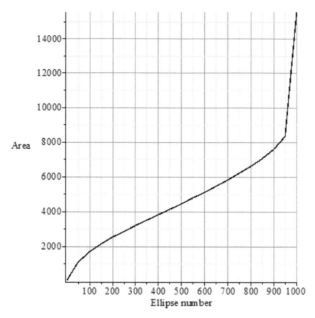

Figure 1.19   Area of Petunin ellipses covering 1000 exponentially distributed points sorted by radius – test 1.

worst achieved result is 0.7% of tested points outside, so only several testing points were not inside the ellipse. Exponential distribution can result in huge coordinate values, so all points were scaled into squares 100 by 100.

Figures 1.19 and 1.20 describe the area of ellipses, created according to basic points – from inner ellipse to ellipse covering all basic points and most tested points.

These figures show a high density of points located near the corner (zero coordinates) and medium impact on ellipse areas from these points, and several points far from the corner, each adding a bigger ellipse than the previous one. So when the exponential distribution is covered with Petunin ellipsoid, rare outer points result in bigger ellipses.

1.6 Conclusion

The proposed algorithm for constructing optimal Petunin ellipsoid in design space is effective and statistically robust. It allows decreasing the amount of points in design space to reduce ellipse areas and keep most points inside. This algorithm demonstrates robustness to outliers, a useful possibility to arrange the point on the probability to cover them by the ellipsoid.

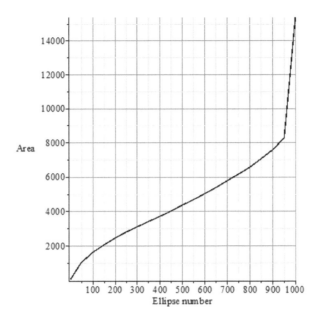

Figure 1.20 Area of Petunin ellipses covering 1000 exponentially distributed points sorted by radius – test 2.

It has a theoretically precise confidence level, which is demonstrated using statistical properties of points inside for five different distributions, each tested with two parameter combinations – 100 tests for each combination. The experiments confirmed the theoretical properties of the Petunin ellipses.

The Petunin ellipsoid is a useful tool for arranging multivariate data and finding anomalies using statistical depth. Thus, future research directions within the topic domain are developing computationally effective algorithms for constructing high-dimensional conformal prediction sets using genetic and other algorithms and inventing new multivariate statistical tests for the homogeneity of multivariate samples in high-dimensional spaces.

References

[1] V. Barnett, 'The ordering of multivariate data', Journal of the Royal Statistical Society. Series A (General), vol. 139, pp. 318–355, 1976

[2] J. W. Tukey, 'Mathematics and the picturing of data', Proc. of the International Congress of Mathematician, Montreal, Canada, pp. 523–531, 1975.

[3] H. Oja, 'Descriptive statistics for multivariate distributions', Statistics and Probability Letters, vol. 1, pp. 327–332, 1983.

[4] R. J. Liu, 'On a notion of data depth based on random simplices', Annals of Statistics, vol. 18, pp. 405–414, 1990.

[5] G. Koshevoy, K. Mosler, K. 'Zonoid trimming for multivariate distributions', Annals of Statistics, vol. 25, pp. 1998–2017, 1997.

[6] P. J. Rousseeuw, M. Hubert. Regression depth. Journal of the American Statistical Association, 94(446):388–402,1999.

[7] I. Cascos, 'Depth function as based of a number of observation of a random vector', Working Paper 07-29, Statistic and Econometric Series 07, vol. 2, pp. 1–28, 2007.

[8] T. I. Lange, P. F. Mozharovsky, 'Determination of the depth for multivariate data sets', Inductive simulation of complex systems, vol. 2, pp. 101–119, 2010.

[9] Y. Zuo, R. Serfling, 'General notions of statistical depth function', Annals of Statistics, vol. 28, pp. 461–482, 2000.

[10] S. I. Lyashko, D.A. Klyushin, V. V. Alexeyenko. 'Multivariate ranking using elliptical peeling', Cybernetics and System Analisys, 49, pp. 511–516, 2013.

[11] K. Mosler, P. Mozharovskyi, 'Choosing among notions of multivariate depth statistics', arXiv:2004.01927v4, 2021.

[12] S. S. Wilks, 'Statistical Prediction With Special Reference to the Problem of Tolerance Limits', Annals of Mathematical Statistics, vol. 13, pp. 400–409, 1942.

[13] A. Wald, 'An extension of Wilks method for setting tolerance limits', The Annals of Mathematical Statistics, vol. 14, pp. 45–55, 1943.

[14] J. Tukey, 'Nonparametric estimation. II. Statistical equivalent blocks and multivarate tolerance regions', The Annals of Mathematical Statistics, vol. 18, pp. 529–539, 1947.

[15] D.A.S. Fraser, I. Guttman, 'Tolerance regions', The Annals of Mathematical Statistics, vol. 27, pp. 162–179, 1956.

[16] I. Guttman, 'Statistical Tolerance Regions: Classical and Bayesian Griffin', Hartigan, J., editor. London.; Clustering Algorithms John Wiley; New York: 1970.

[17] J. Aichison, I. R. Dunsmore, 'Statistical Prediction Analysis' Cambridge University Press, 1975.

[18] I. G. Bairamov, Yu. I. 'Invariant confidence intervals for the main mass of values from the distribution in a population', J Math Sci, vol. 66, pp. 2534–2538, 1993.

[19] G. Shafer, V. A. Vovk, 'A Tutorial on Conformal Prediction', Journal of Machine Learning Research, vol. 9, pp. 371–421, 2008.

[20] J. Li, R. Liu, 'Multivariate spacings based on data depth: I. construction of nonparametric multivariate tolerance regions', The Annals of Statistics. 36, pp. 1299–1323, 2008.

[21] J. Lei, J. Robins, L. Wasserman, 'Distribution Free Prediction Sets', Journal of the American Statistical Association, vol. 108, pp. 278–287, 2013.

[22] E. Ndiaye, I. Takeuchi, 'Computing Full Conformal Prediction Set with Approximate Homotopy', NeurIPS Proceedings, pp. 1384–1393, 2019.

[23] Yu. Petunin, B. Rublev, 'Pattern recognition with the help of quadratic discriminant function', J. Math. Sci., vol. 97, pp. 3959–3967, 1999.

[24] B.M. Hill, 'Posterior distribution of percentiles: Bayes' theorem for sampling from a population', Journal of the American Statistical Association, 63, pp. 677–691, 1968.

2

Data Analysis on Automation of Purging with IoT in FRAMO Cargo Pump

R. Prasanna Kumar[1] and V. Ajantha Devi[2]

[1]Indian Maritime University – Chennai Campus, India
[2]AP3 Solutions, India

Email: Prasanna.r@imu.ac.in; ap3solutionsresearch@gmail.com

Abstract

Oil and chemical tankers use the FRAMO system to transport cargo from ship to shore. FRAMO cargo pumps, which are installed inside cargo tanks, are an essential component of the system. The results of the purging operation are used to determine whether or not these cargo pumps are operationally ready. The purging procedures confirm the integrity of the sealing arrangements on both the cargo and hydraulic systems. The cofferdam that separates the cargo from the hydraulic fluid collects any leakage from the FRAMO cargo pump's seals. The AUDRINO control board and electronic control of various solenoid-signaled hydraulic actuated valves are recommended for this auto-mated purging procedure. Control signals from the shore control center or the inbuilt timer circuit can be delivered via IoT. This control board also oversees the purging sequence. The leak-off liquid is lifted from the cofferdam space to the sample container by automated purging.

The identification of the liquid is essential for obtaining the purging result. The method proposed in this paper employs three distinct sensors to identify the liquid in the cofferdam space in the autonomous ship environ-ment. The three parameters are density meter, pH meter, and a color sensor. These three characteristics distinguish the cargo liquid from the hydraulic oil used in the system. This test result is useful for cargo operation plan-ning. The content received in the cofferdam is revealed by comparing the measured data set to the preloaded database. The major goal of this research is to use physicochemical data to predict oil content. Two distinct data sets were obtained in this investigation. These data sets include three major cargo and hydraulic oil physicochemical properties. Using the random forests algorithm, the instances were effectively identified as cargo oil or hydraulic oil with an accuracy of 98.6229%. The detection of both cargo oil and hydraulic oil was then classified using three distinct data mining techniques: k-nearest-neighbor, support vector machines, and random forests. The random forests algorithm provided the best accurate classification.

2.1 Introduction

The energy requirements of today's industrial activity are critical in any part of the world. A variety of products derived from crude oil meet this energy demand. Product tankers are the preferred mode of transportation for these various oils. Product tankers transport a wide range of liquid cargoes, including various chemicals used by different industries. The time it takes to transport cargo from ship to shore is used to determine these tankers' operational efficiency. Because product tankers transport a wide range of commodities, the FRAMO [1] method is widely used for cargo operations. The readiness of the FRAMO cargo pump is critical for optimizing cargo discharge time. A purging test, which analyses the amount of leaked liquid and the type of liquid in the cofferdam, certifies the cargo pump's readiness. This article describes the automatic purging operational test in the context of autonomous shipping [2–6].

2.2 FRAMO Cargo Pump

The FRAMO cargo pump (Figure 2.1) is housed within the cargo tank. The impeller of this centrifugal pump is located at the bottom of the tank. The hydraulic motor, which is connected to this impeller, is also located at the bottom of the tank. Pressurized hydraulic oil required to power this pump is delivered to the motor via pipe stock. This pipe stock takes up nearly the entire height of the tank. The hydraulic system's required pressure is generated at Powerpack, which is located in the remote Engine Room.

2.2.1 Cofferdam

These pumps were built with a cofferdam [7] area between the cargo and hydraulic seals. The hydraulic motor is installed inside the tank, and hydraulic oil is allowed to flow into the tank to power it. The seal is installed between

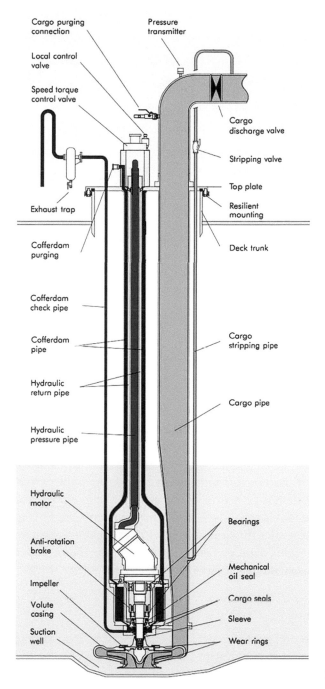

Figure 2.1 FRAMO pump.

Figure 2.2   Arduino board.

the shaft and casing to prevent hydraulic oil leakage through the driving shaft. While cargo oil is being lifted through the pump impeller, a cargo seal is located between the shaft and the pump casing to prevent cargo leakage around the shaft. A cofferdam space exists between these two seals to catch any leaks from either side.

2.2.2 Cargo pump purging operation

The condition of the cargo pump is critical for an efficient cargo transfer operation [8]. If any deficiencies in the cargo pump conditions are discovered, operators implement corrective plans for alternative transfer methods such as the use of a spare pump. To avoid the vessel turnaround delay, this planning should have been completed prior to the vessel berthing in the terminal.

A purging test certifies the cargo pump's readiness. The setup is available at the tank's top to allow purging air/nitrogen to pass through the cofferdam space. When the cofferdam is pressurized, the leakage content collected in this space is lifted to the tank's top. The leakage liquid is collected by an exhaust trap installed in the outlet pipeline. Identifying the leakage content aids in determining which seal is damaged, and measuring the volume of the liquid reveals the extent of seal damage.

2.3 ARDUINO Board

Arduino [9] is a hardware and software which is customizable electronic platform, as in Figure 2.2. It can read various sensor inputs and respond to any IoT signals or inbuilt activation signals. In addition, the output signal from this board can perform the physical activities as programmed in this control board.

Arduino board is used in various smart applications such as simple home automation to intelligent cargo management systems. Cloud enabled feature of this board very much useful on data handling on any application.

2.3.1 Fork-type density meter

Density is one of the important characteristics useful to identify the liquid. This project chooses a liquid density measurement sensor that uses a tuning fork [10]. Tuning fork sensor works based on the resonance principle to measure the density of liquids. Being a quasi-digital sensor, it can directly monitor liquid density. Even though the measuring sensor of lightweight and simple in structure, its precision and reliability levels are high.

The tuning fork liquid density measurement sensor inserts the vibrating element in the measuring chamber filled with liquid, that is operated in a piezoelectric manner. The actuator allows the tuning fork to vibrate at its natural frequency. The frequency of vibration is inversely proportional to the density of the surrounding liquid. The extra mass of the tuning fork varies when it comes into contact with the liquid being measured, resulting in changes in vibration frequency (vibration cycle). The detector detects the vibration frequency by picking up the vibration signal.

The temperature sensor is also added to the part of this density meter as the density of liquids varies depending upon the temperature. The density of the liquid being examined can be calculated by measuring changes in the natural frequency or vibration cycle and its temperature value.

2.3.2 pH Meter

pH value is a measure of hydrogen ions [11, 12] (acidic or alkalinity) in a solution. This is measured electronically in a pH meter. A pH meter is made up of a voltmeter that is connected to two electrodes, one is a pH-responsive electrode, and another is the reference (unvarying) electrode. The two electrodes work similarly to a battery when immersed in a solution.

The glass membrane and the requisite reference electrodes are usually included in the same electrode body, which is the most typical design for pH electrodes. The inner fill solution touching the glass membrane in this configuration has a fixed H+ activity, as shown in Figure 2.3. This inner solution typically has an Ag/AgCl reference electrode in contact with it, and the solution includes 0.1 M HCl saturated with AgCl. A second Ag/AgCl reference

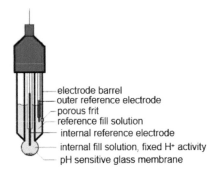

electrode barrel
outer reference electrode
porous frit
reference fill solution
internal reference electrode
internal fill solution, fixed H+ activity
pH sensitive glass membrane

Figure 2.3   Glass electrode pH meter.

electrode is positioned in the next compartment encircling the inner solution. Through a porous frit on the side of the electrode barrel, the measuring solution comes into touch with an external solution with unknown H+ activity.

As the internal fill solution pH is fixed, the variation in the potential between the two electrodes is due to the pH of the solution in contact with the second compartment. This digital device which can read out the difference in potential in terms of pH.

2.3.3 Color sensor

Liquid color can be used as one attribute to identify the leakage content. To find the color, the light is passed through the sample to the color sensor [13, 14]. The color sensor is a configurable silicon photodiode that converts current to frequency. This module has three color filters. The basic theory is various colors are a mixer of three basic colors in different proportions. When one color filter is selected, it allows only one color and blocks the other two; for example, if the red color filter is selected, it allows red but blocks blue and green. Basis of the intensity of the red color its equivalent pulses of certain frequency delivered in output. Similarly, two other color intensities can also be obtained. In the end, three different frequency values obtained. The color can be identified by analyzing these three frequency values.

2.3.4 Level sensor

The liquid level in the sample container is measured to calculate the leakage content volume. The capacitance type level measurement is a good choice to perform this task. When two electrodes [15] with known distance are inserted in the liquid column. The measured capacitance value changes depending upon the die-electric constant of the material between the two electrodes.

As air is a nonconductor, capacitance measured in this circuit is the basis on the liquid level between the two electrodes. Vessel trim and list influence the level measurement, so the necessary correction value is to be added. With the known area of the sample container volume of the liquid can be calculated.

2.3.5 Internet of Things (IoT)

Internet of Things is a network connection of various physical objects [16, 17] through internet that have been embedded with programmable actions such as sensing [18], data collection, data processing [19], and data/signal transmission [18]. It is the connection point between the physical and digital worlds. Much development has taken place on IoT maritime applications [20–22], particularly related to cargo carriage. As autonomous ships evolve, more such applications are in the developing stage of monitoring vessel machinery [23] and operations.

2.4 Proposed Method

The proposed method of automating the purging operation can be explained with three suboperations, as in Figure 2.4.

1. Purging Automation

2. Data Acquisition and Analysis

3. Communication

2.4.1 Purging automation

An air/nitrogen supply valve at the top of the tank can be used to manually execute the purging. When automating the process, this valve must be changed to signal control. The Arduino control board can be programmed with the activation signal. The FRAMO hydraulic system [1] itself can provide the external power required to operate these valves. This operation is carried out using hydraulically powered and signal-controlled valves.

The presence of moisture in the air used affects the test findings. The service air must be drained to ensure dryness for accurate data collection. Place the auto-drain valve in the service air system and turn it on before opening the air supply valve removes the moisture. The timer feature on the Arduino board [9] can be used to program the entire operation. Even though some air is lost during each purging function, this procedure ensures that

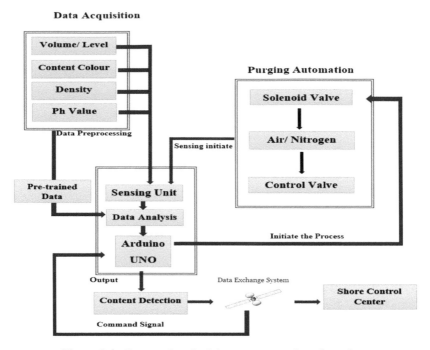

Figure 2.4 Proposed methodology on automation of purging.

good-quality air is utilized. The collected leakage liquid in the exhaust trap should be emptied into the sample beaker at the completion of purging the cofferdam. Another auto-operated drain valve should be installed below the exhaust trap for this purpose.

The cleanliness of the sample beaker has an impact on the test findings; to get the right result, each purging procedure requires a fresh sample beaker. Therefore, at the start of the voyage, several samplers must be stored upside down, and at the end of each purging cycle, these sample beakers are positioned upright below the drain valve. This motion is carried out by a robotized arm that is controlled by an Arduino system.

2.4.2 Data acquisition and analysis

The following sensors are used to capture the physical parameters of the leaked liquid. Color sensor [13, 14], density meter [10], and pH meter [11, 12]. The level of the leaking liquid is detected using a level sensor. The sample beaker must travel to different testing stations using the planned conveyor system, and these sensors must be set in the correct order. After all, testing has been completed, and the sample beaker and leakage liquid should be placed into the garbage can. The same Arduino board transfers the sample to different test points and waits for the sensors to collect its sensing value.

The level sensor and data analysis of these physical attributes allow us to confirm the leaked liquid and the severity of the seal damage.

2.4.3 Communication

This proposal's communication is split into two parts: internal and external. Different automatic valves, various sensors, conveyor movement, and the central control module are all connected via internal communication. Optical fiber cables are used to transmit signals between various components. The Arduino board's supervisory role is efficiently delivered through efficient data transmission throughout the ship. The satellite facility is used for external communication. The VHF data exchange system (VDES) [24] allows for efficient data transmission between the ship and the shore control center.

2.5 Experimental Data Analysis

The development of new sensory materials is critical to modern IoT sensing and intelligent system. These novel materials aid in developing modern sensing technologies [25]. The study of chemometrics, a branch of chemistry that employs analytical data to develop appropriate measuring methods for providing relevant chemical information. The algorithm is being developed in this direction in order to reach a conclusion in this investigation. The following are a few of the factors that played a role in selecting sensors and developing the algorithm.

(i) Without compromising data that would somehow or another be important for future arrangement, the entire arrangement of qualities that describe a dataset of detecting examinations can be utilized as a contribution to multidimensional projection calculations.

(ii) Other multidimensional perception strategies, like equal directions, empower the recognizable proof of the viewpoints that most altogether add to the sensor's separating limit.

(iii) Nonlinear models, for example, have been found to be beneficial for managing bio-sensing data in particular situations.

There are 500 instances in the dataset, each with 15 attributes for hydraulic oil and cargo products. Objective tests (such as pH values, density, and color) are used as inputs, and the aquation is based on sensory data.

Table 2.1   The physicochemical properties of hydraulic oil.

Hydraulic oil	pH value	Color (appearance)	Density
Castrol hyspinAWH32	9.00	Golden	870

The data set's purpose is to use various physicochemical parameters [25] to determine the content sample of the leakage from the cofferdam. Only physicochemical and sensory variables are used due to privacy and logistical concerns. The k-fold cross-validation mode [26] and the percentage split test modes are employed for evaluation.

This investigation randomly divides the database into two unique data-sets for percentage split mode. The training set is the main data from which the data mining system attempts to extract knowledge. The extracted data can be compared to the second set, referred to as the testing set.

Different k values are examined for each technique in k-fold cross-validation mode. When the k value is set to 10, the best classification results of each technique are achieved. Using the 10-fold cross-validation approach, the training data contains 80% of the dataset at random, whereas the second set contains 20% of the testing data. The main three physicochemical properties of hydraulic oil and few cargo product sets are presented in Tables 2.1 and 2.2, respectively.

Datasets were developed in the original form of these datasets, combining hydraulic and cargo products. The objective is to sort container samples into two categories: hydraulic oil and cargo product. **Our research** used three different data mining methods [25,26]. Support vector machines (SVM), K-nearest neighbor (k-NN), and random forests (RF) were the categorization techniques used on the data set.

1.   Classifiers for k-Nearest Neighbor: The fundamental presumption behind the KNN classifier is that assuming most of a question test's K most-comparable examples have a place with a particular classification, then, at that point, the inquiry test does also. KNN does not require any earlier information.

2.   Random Forests: RF produces another preparation set by arbitrarily choosing tests with situations from the first preparing set, then, at that point, rehashes the previously mentioned techniques to prepare various choice trees to develop an arbitrary timberland. Every choice tree is

Table 2.2 The physicochemical properties of cargo oil.

Cargo products	pH value	Color (appearance)	Density
Alkanes (C6-C9)	7.00	Silver	700
Alkyl (C3-C4) benzenes	2.25	Brown	863
Alkylbenzene sulfonic acid, sodium salt solution	8.00	Colorless	1070
Ammonium polyphosphate solution	6.00	White	1900
Ammonium sulfate solution	5.50	White	1770
n-Amyl alcohol	7.00	Colorless	814
Calcium hydroxide slurry	11.27	Colorless	2211
Calcium lignosulfonate solutions	3.00	Yellow	1081
Camelina oil	7.50	Bright yellow	840
Cashew nutshell oil (untreated)	4.50	Dark reddish-brown	1009
Castor oil	6.34	Pale yellow	959
Choline chloride solutions	6.50	White	1100
Citric acid (70% or less)	4.50	White	1660
Coal tar	5.30	Black	960
Cocoa butter	5.60	Pale yellow	920
Coconut oil	5.00	White	910
Coconut oil fatty acid	6.00	Colorless	925
Corn oil	7.00	Dark yellow	910
Cotton seed oil	3.50	Light golden	917
Cycloheptane	7.20	Colorless	751

given a question test and is utilized to settle on a choice prior to casting a ballot to figure out which class it has a place with [27].

3. Support vector machines (SVMs): To arrange information in high-dimensional space, SVM searches for hyperplanes. SVM's motivation is to augment the edge among hyperplanes and support vectors, which can be refined by changing over the undertaking into a raised quadratic programming issue [28].

Moreover, some standard presentation estimations, for example, precision, Recall, *F* measure, and ROC, are determined to assess the calculations' exhibition. For both test modes, the random forests algorithm delivered the best order results for water powered oil and freight item tests. With this strategy, cross-approval and rate split mode precision is 98.622% and 98.461%, individually. Table 2.3 obviously shows that the random forests calculation outperforms the other two calculations in both assessment modes.

Table 2.3   The random forests algorithm.

Evaluation modes	Classifier	Precision	Recall	*F* measure	ROC
Cross-validation	SVM	98.2	98.2	98.2	97.6
	k-NN	98.3	98.3	98.3	98.0
	RF	98.6	98.6	98.6	98.8
Percentage split	SVM	98.3	98.3	98.3	97.6
	k-NN	98.3	98.3	98.3	97.7
	RF	98.6	98.6	98.6	98.9

2.6 Limitations

Current investigation considered the following limitations

(a) Vessel carrying a single cargo – In practice, tanker vessels can transport multiple cargoes in different tanks. In this regard, the presented algorithm is only useful if the cargo loaded in the tank is specified. Multi-cargo information will be added as a new data set, and the algorithm will need to be fine-tuned using multi-layer classifications.

(b) Only one seal is leaking – The algorithm is now applied with the assumption that only one seal is leaking and that only a specific liquid, either cargo or hydraulic oil, is collected in the cofferdam. If both seals fail simultaneously, the sensory values of density and pH cannot determine the leakage liquid. In the case of out-of-range property values, a new algorithm must be developed to determine whether the out-of-range value is the result of multiple leaks or a lack of sensor accuracy.

2.7 Conclusion

In the present installed rehearses, the total cleansing method is performed physically. The compressed air supply through the cofferdam powerfully eliminates the substance in that space, and it helps the team part to recognize any disappointment of seals and the degree to which the disappointment is. This paper examined robotizing the cleansing system under automated climate and imparting the analysis results to the control community for the decision utilizing the specific pump [29].

Classifier analysis on hydraulic oil and shipping product data sets is covered in this study. The results are provided as a proportion of correctly identified instances, recall, precision, *F* measure, and ROC after cross-validation or percentage split mode.

Various classifiers, such as k-nearest neighbor, support vector machines, and random forests, are tested on datasets. The random forests (RF) classifier surpasses the support vector machine and the k-nearest neighbor algorithm in classification tasks, according to the results of the trials.

References

[1] www.framo.com
[2] Maritime, U.K., 2018. Maritime autonomous surface ships - UK code of practice. Retrieved from. https://www.maritimeuk.org/documents/305/MUK_COP_2018_V2_ B8rlgDb.pdf.4
[3] Fan, C., Wróbel, K., Montewka, J., Gil, M., Wan, C., Zhang, D., 2020. A framework to identify factors influencing navigational risk for Maritime Autonomous Surface Ships. Ocean. Eng. 202, 107188.
[4] S. Garg et.al, "Autonomous Ship Navigation System", in 2013 Texas Instruments India Educators' Conference, pp. 300–305
[5] D. Rodriquez, M. Franklin and C. Byrne, "A Study of the Feasibility of Autonomous Surface Vehicles," Worcester Polytechnic Institute, Dec.2012.
[6] F. Fahimi, "Autonomous Surface Vessels," in Autonomous Robots: Modeling, Path Planning, and Control, 1st ed.: Springer, 2008, ch. 7, pp.221–260.
[7] Krikkis, Rizos N. "A thermodynamic and heat transfer model for LNG ageing during ship transportation. Towards an efficient boil-off gas management." Cryogenics 92 (2018): 76–83.
[8] Sumarno, P. S., Dwi Prasetyo, and Saiful Hadi Prasetyo. "Identifikasi Penyebab Kerusakan Seal Cargo Pump Dalam Proses Discharging Muatan Kimia Cair." Dinamika Bahari 8, no. 2 (2018): 2045–2062.
[9] Auråen, Jonas. "Low-cost CTD Instrument-Arduino based CTD for autonomous measurement platform." Master's thesis, 2019.
[10] Megantoro, Prisma, Andrei Widjanarko, Robbi Rahim, Kunal Kunal, and Afif Zuhri Arfianto. "The design of digital liquid density meter based on Arduino." Journal of Robotics and Control (JRC) 1, no. 1 (2020): 1–6.
[11] Patil, Amruta, Mrinal Bachute, and Ketan Kotecha. "Identification and Classification of the Tea Samples by Using Sensory Mechanism and Arduino UNO." Inventions 6, no. 4 (2021): 94.
[12] Ariswati, Her Gumiwang, and Dyah Titisari. "Effect of temperature on ph meter based on arduino uno with internal calibration." Journal of Electronics, Electromedical Engineering, and Medical Informatics 2, no. 1 (2020): 23–27.

[13] Nazeer, A. Mohamed, S. Sasikala, A. Kamalabharathy, SD Kirubha Dharshni, M. Nandhini Lakshmi, and K. SherinPriya Dharshini. "Colour Sensing Using Robotic ARM." (2020).

[14] San Hlaing, Ni Ni, Hay Man Oo, and Thin Thin Oo. "Colour Detector and Separator Based on Microcontroller." (2019).

[15] Megantoro, Prisma, Andrei Widjanarko, Robbi Rahim, Kunal Kunal, and Afif Zuhri Arfianto. "The design of digital liquid density meter based on Arduino." Journal of Robotics and Control (JRC) 1, no. 1 (2020): 1–6.

[16] Statheros, T., Howells, G., Maier, K.M., 2008. Autonomous ship collision avoidance navigation concepts, technologies and techniques. J. Navig. 61 (1), 129–142.

[17] DNV GL. 2018. "Remote-controlled and Autonomous Ships in the Maritime Industry." Group Technology and Research, Position Paper 2018.

[18] Richard Tynan, Gregory MP O'Hare, and Antonio Ruzzelli. Autonomic wireless sensor network topology control. In 2007 IEEE International Conference on Networking, Sensing and Control, pages 7–13. IEEE, 2007.

[19] Yi Xu and Abdelsalam Helal. Scalable cloud–sensor architecture for the internet of things. IEEE Internet of Things Journal, 3(3):285–298, 2016.

[20] John A Stankovic. Research directions for the internet of things. IEEE Internet of Things Journal, 1(1):3–9, 2014.

[21] Etzioni, A., and O. Etzioni. 2017. "Incorporating Ethics into Artificial Intelligence." Journal of Ethics 21 (4): 403–418.

[22] Komianos, A. 2018. "The Autonomous Shipping Era.: Operational, Regulatory, Quality Challenges." International Journal on Marine Navigation and Safety of Sea Transportation 12 (2): 335–348. doi:10.12716/1001.12.02.15.

[23] Kongsberg Maritime AS. 2018. "Autonomous Ship Project." https://www.km.kongsberg.com

[24] Lázaro, Francisco, Ronald Raulefs, Wei Wang, Federico Clazzer, and Simon Plass. " Data Exchange System (VDES): an enabling technology for maritime communications." CEAS space Journal 11, no. 1 (2019): 55–63.

[25] P. Cortez, A. Cerderia, F. Almeida, T. Matos, and J. Reis, "Modelling wine preferences by data mining from physicochemical properties," *In Decision Support Systems, Elsevier*, 47 (4): 547–553. ISSN: 0167-9236.

[26] J. Han, M. Kamber, and J. Pei, "Classification: Basic Concepts," in *Data Mining Concepts and Techniques*, 3rd ed., Waltham, MA, USA: Morgan Kaufmann, 2012, pp. 327–393.

[27] W. L. Martinez, A. R. Martinez, "Supervised Learning" in *Computational Statistics Handbook with MATLAB*, 2nd ed., Boca Raton, FL, USA: Chapman & Hall/CRC, 2007, pp. 363–431.

[28] J. Han, M. Kamber, and J. Pei, "Classification: Advanced Methods," in *Data Mining Concepts and Techniques*, 3rd ed., Waltham, MA, USA: Morgan Kaufmann, 2012, pp. 393–443.

[29] R. Prasanna Kumar, Dr. V Ajantha Devi, "Automation of FRAMO Cargo Pump purging with IoT" in 21st Annual General Assembly of the International Association of Maritime Universities Conference, IAMU AGA 2021, Proceeding of the International Association of Maritime Universities (IAMU) Conference-Arab Academy of Science, Technology and Maritime Transport, Alexandria, Egypt, 2021, pp. 444–453.

3

Big Data Analytics in Healthcare Sector: Potential Strength and Challenges

Swati Gupta[1], Meenu Vijarania[1], Akshat Agarwal[2], Aarti Yadav[1], Ranadheer Reddy Mandadi[3], and Sunil Panday[4]

[1]Department of Computer Science & Engineering, School of Engineering and Technology , K R Mangalam University, Gurugram, India
[2]Department of Computer Science & Engineering, Amity School of Engineering and Technology, Amity University Haryana, Gurugram, India
[3]Department of Remote Sensing & Geographic Information Systems, Asian Institute of Technology, Klong Luang, Pathum Thani, Thailand
[4]Department of Mathematics, National Institute of Technology Manipur, Langol India

Email: swati@krmangalam.edu.in; meenu@krmangalam.edu.in; akshatag20@gmail.com; aartighumi@gmail.com; Ranadheer@ait.asia; sunilpanday@hotmail.co.in

Abstract

In current era with rapid advancement in data acquisition and sensing technologies healthcare institute and hospitals have started collecting enormous data about their patients. The issue with healthcare data is that it has an unprecedented diversity in terms of its data formats, data types, and the rate at which it needs be analysed to deliver the required data. This variety is not just limited by its amount. Given the variety and number of sources that really are constantly expanding, it is now difficult to handle such vast volume of unstructured data using conventional tools and methodologics. As a result, sophisticated analytical tools and technologies are required, especially big data analytics techniques, for interpreting and managing the data related to healthcare.

   Big data analytics can be used in healthcare for better diagnosis, disease prevention, telemedicine (especially when using real-time alerts for immediate care), monitoring patients at home, avoiding hospital visits, integrating medical imaging for a wider diagnosis. Hence, Healthcare data analytics have gained a lot of attention, but their pragmatic application has yet to be adequately explored. The chapter aim to cover various problems and potential solutions that frequently arise when big data analytics are utilized in medical institutions. We have reviewed big data analytical tools and methods in healthcare such as classification, clustering, artificial neural network and fuzzy.

3.1 Introduction

The way the "Big data "is defined nowadays has been changing constantly. The term "big data" refers to a set of collection of large volume of data whose speed, size, complexity needed a new software mechanism that can effectively store and analyse the data [1-3]. The healthcare industry is a crucial sector which has shown that how three V related to data namely variety, velocity and volume are mapped [4].The recent progress in technology led to the ability of collecting enormous data in healthcare sector. As per the recent advancement in Big Data it is suggested that the potential of generating the health care data has increased to 300 billion [5].With the advancement in data acquisition and sensing technologies the vast amount of data related to their patients has been collected by hospitals and several healthcare institutes. However, despite the collection of large amount of data it lacks its proposer utilization. Consequently, effective analytical approaches are useful in turning the provided data into some meaningful information in order to build knowledge from the healthcare data. Thus, data analytics has become the critical component in transforming the healthcare data from being reactive into proactive form. This impact of analysing the healthcare data has been increasing in the coming years. In order to enable clinicians create a unique profile of each patient that can precisely calculate the risk that a patient will have any medical ailment in the near future, healthcare data analysis extracts patterns that are hidden in the data set. [6].

Healthcare data are collected from wide range of sources such as images, sensors, biomedical literature and electronic records. The heterogeneity of data collection and its representation causes several challenges in performing pre-processing and analysis of the data in different perspective [7]. The "data privacy gap" is another challenge that medical researchers and computer scientists encountered as the data related to healthcare data is a sensitive information since it can reveal certain compromising information regarding an individual [8]. Despite of several challenges that exist in handling the big

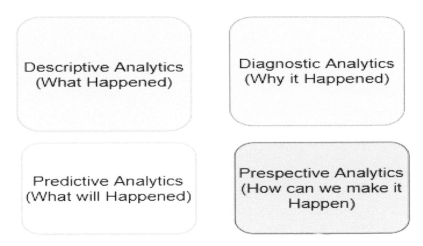

Figure 3.1 Different perspectives for big data healthcare analytics.

data analytics in healthcare it provides potential benefits to enhance the quality of a healthcare system by understanding different dimension as shown in Figure 3.1 which are discussed as follows:

- Descriptive analytics - Is utilised to comprehend the past and current decisions in healthcare which is able to find out the answer what has happened. It is used for creating and visualizing the reports containing information about patient hospitalization, physician information etc.

- Diagnostic analytics- Knowledge is utilized to find out new inventions it provide answer to the possible reasons why it happened.

- Predictive analytics- It focuses on what will occur in the future by analysing the historical healthcare data. It finds out the relationship that exist among the data sets so as to optimize the healthcare resources.

- Prescriptive analytics— determines how the things can make it happen it uses medical knowledge.

The primary sources from where healthcare data are collected is shown in Figure 3. 2. The data is processed by different organization to form a big data chain [9] .The various sources of health care data is discussed below.

- Electronic Health Records (EHR)
  It contain digitized record that comprises of complete set of data that are related to a particular patient. EHRs can be accessed and edited by authorized users. It is useful in improving the quality and convenience of patients [10].

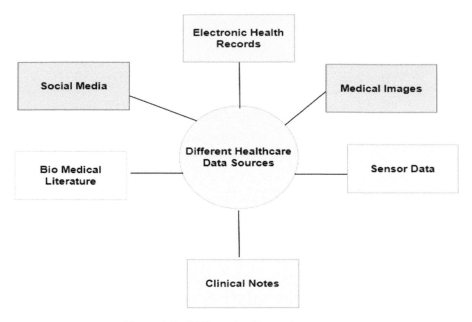

Figure 3.2   Different healthcare data sources.

- Medical Images
  Medical Images are useful for the researchers and clinicians as it can help in disease monitoring and planning required during the treatment process. Inferences are drawn from the images that can provide clear picture of the patient medical condition [11].

- Sensor Data
  Sensor data has created a meaningful impact in the healthcare. However, it faces the challenges of overloading. Hence, an efficient analytical techniques are required to convert data into meaningful knowledge [12].

- Clinical Notes
  Clinical notes are unstructured data format that contain clinical information. It is quite challenging to analyse the information present in these notes due to the difficulty associated with it in transforming clinical text notes into structured format [13].

- Bio-Medical Literature
  The research in healthcare depends on Bio medical literature. Several text mining methods are used for providing preservation, usability and accessibility that relies on applying the knowledge discovery methods in the medical field [14].

- Media
  With the rise of social media platforms like Twitter, blogs, forums, and online groups, a tonne of information about many facets of healthcare is produced. On this data several data mining techniques can be used to extract some useful pattern and knowledge that are used to make important inferences [14].

3.2 Literature Review

The prior efforts that are carried out in Big data analytics in healthcare sector are discussed in detail which are as follows:

Zhang et al. [15] discussed that a healthcare is considered as a system that comprises of three fundamental parts to provide medical care services the core providers that include physicians, nurses ,technicians and hospital administration, critical care that deals with medical care such as insurance and medical research and the beneficiaries that include patients and public. Bates et al. [16] presented different examples of patients and specifies different method of reducing risk by using big data analytics. The author mentioned the data required, infrastructure and machine learning algorithm needed to gain learning from previous experience. Driscoll et al. [17] discussed overview of big data technologies in cloud computing which are used for handling bigger datasets for example in sequencing human genomes which are used for understanding genetic variations.

Quinn et al. [18] discusses about the evaluation of the smartphone based diabetic management software by using a statistical model, relevant data, behavior and trends ,provided tools which provide user advise on diabetics for efficient management of the disease. Wamba et al. [19] creates a diseases risk profile for each patient and biomedical discovery. Patient profile comprises of electronic medical records, patient history and information related to biological factor which is helpful in developing a personalized plan. Andreu-Perez et al. [20] discussed the improvement achieved in health industry because of using big data.It further explain the potential benefit that exist for different sources of data such as biomedical, sensor that are represented in both structured and unstructured format.

Hilario et al. [21] provides an investigation on data analytics that uses a large amount of data from different biological specimens like DNA.It then discovers pattern among pathological states by using classification algorithm which helps in discovering the pattern. Murdoch et al. [22] discussed the different applications of applying big data in health care that uses economic framework which helps in enhancing the quality of health care sector. Barrett et al. [23] proposed a model based on bayesian forecasting algorithms that helps in developing drug–specific dashboards that is used for performing better decisions and provide necessary education to the patient . It reduces the medication errors, decreases the length of days the patient is admitted to hospital .The different data visualization tools helps in displaying the patient profile from its electronic medical records that specifies lab values and biomarker.

Gligorijevic et al. [24] discusses the different integrative methods in big data that are useful in identifying biomarkers discovery and diseases subtyping . Kuiler et al. [25] developed a conceptual framework for analysing the data. The semantic challenges in big data are addressed by IT-supported ontology based approaches. It developed the different conditions for different artefacts such as lexicon, ontology etc. Bardhan et al. [26] proposed a new model that helps in predicting readmission of patients that are suffering from congestive heart failure. The model is useful in tracking patient information by analysing the data obtained from 67 hospitals over the 4 years that contain clinical, demographic and administrative data. Basole et al. [27] uses visual analytic tool that helps in analysing the healthcare data related to asthmatic in children patients. Ajorlou et al. [28] developed bayesian model based on linear prediction which indicates risk adjustment according to condition of the patient. The data is collected from 82,000 patients in a year are evaluated in health administration. Delen et al. [29] make use of data mining techniques namely neural networks, decision trees and support vector machines to build on predictive models to analyse the survival rate in prostate cancer. A total of 120,000 records were collected from different sources such as surveillance resulting in developing 77 variables used to draw statistical inference.

Liu et al. [30] proposed an agent-based simulation model that helps in studying the different responsible based organisations. The life-threatening elements are identified in developing the payment design model. The healthcare analytical model helps in developing an efficient health policy and further is useful in making better healthcare decisions. Holzinger et al. [31] uses a text mining techniques namely semantic analysis that is beneficial for clinical decision support system. It is useful in providing decision support by analysing biomedical data. Althebyan et al. [32] proposed a monitoring system related to e-healthcare which are useful in targeting the individual for a given geographical area. It helps in decision support system that reduces risk related to patient health. Table 1 discusses the different analytical techniques that are applied on healthcare data sector.

3.3 Research Methodology

The healthcare industry is collecting enormous data by relying on automation process. Clinical data, patient and emotion based data, administration and

Table 3.1 Analysis technique in big data analytics for the healthcare sector.

S. no.	Author and reference	Analytical technique	Discussion
1	Bardhan et al. [26]	Predictive	It uses mathematical algorithms for finding out predictive patterns. A novel model is presented that is useful in predicting the readmission of patients that suffer from congestive heart problem.
2	Dundar et al. [33]	Prescriptive	To obtain balance assignment workload a survey of 2865 patients are performed that helps in developing patient assignment model.
3	Basole et al. [27]	Descriptive	It improves the healthcare quality by using the analytical tool for 5784 patients related to asthmatic problem in children.
4	Abbas et al. [34]	Descriptive	It uses clinical and administrative data for creating cloud based solution that is useful in providing recommendation regarding health insurance plan.
5	Boulos et al. [35]	Predictive	Predictive tools analysis are performed to represent real time health events like outbreak of any medical problem.

cost activity data are the many types of big data that can be used. However, data analysis is required for the transformation of data into capabilities. Our research methodology comprises of three stages namely: Inputs, Processing and Outputs as shown in Figure 3.3

In Stage I inputs data is collected from the review that are derived from electronic database covering the research articles from Scopus and web of sciences. The articles were reviewed that matches the keywords related to healthcare [24]. We obtain almost 4000 articles. The final dataset comprises of papers that contribute towards the use of big data in the healthcare. Stage II comprises of information extraction, tracking and categorization of topics. Further, text classification technique is applied to formulate the dimension with sub-categories. Finally, stage III represent the output of classification method through tabular representation. The Healthcare big data framework is shown in Figure 3.4

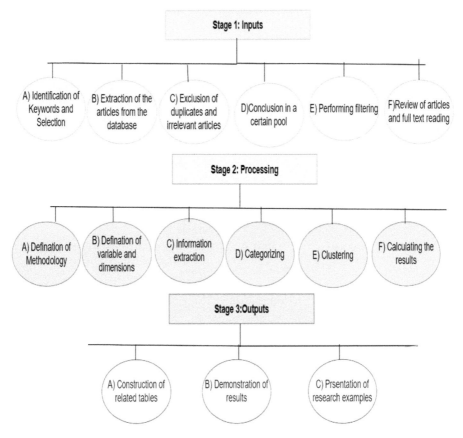

Figure 3.3   Methodology process adopted in data analytics.

Vital elements needed in data analytics used in healthcare are Data Elements, Functional Elements, Human element and Security Element as shown in Figure 3.5.

A.   Data Elements

Data plays a crucial role in analytics to obtain the useful insights. Healthcare related data is analysed for improving the quality of the process. To potentially utilize the data it needs to be obtained from multiple sources. The data collected from multiple sources are diverse in nature sometimes represented in highly structured form while it may be represented as semi-structured form supporting heterogeneity [37]. Hence, handling healthcare data requires efficient data capabilities in terms of data processing and its storage to obtain useful inferences.

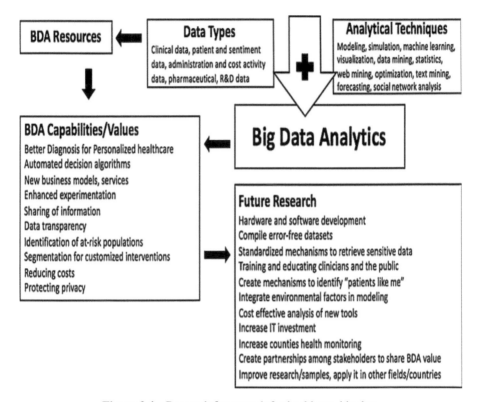

Figure 3.4 Research framework for healthcare big data.

B. **Functional Elements**

To obtain meaningful insights from healthcare data the analysis of the information involves the following elements.

1. Data Preparation and Processing: To provide high quality results the data in this stage are cleaned and formatted. The distributed processing capability is required to process the healthcare data [38].

2. Building Analytical Models: Analytics plays an important role in healthcare sector which are derived from healthcare data. Advanced analytical models are required for obtaining value from healthcare data.

3. Decision-Support Tools: The results must be correlated so that they can be synchronized with existing data. The decision support tools are useful in making decisions.

4. Data Visualization: Decisions are communicated to the decision makers using data visualization tools. The results can be represented in bar charts, graphs and findings using popular tools like tableau.

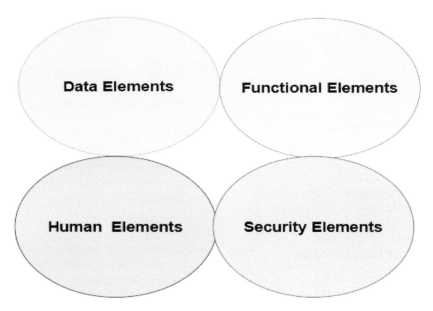

Figure 3.5 Important elements for analysis of healthcare data.

C. Human Element
People are required who can perform the inference of the results obtained by anlyzing the healthcare data [38]. In Healthcare sector there is an urgent need for clinical experts that are responsible for evaluating the output of analytical tools.

D. Security Elements
The elements that are related to security are:

- Data Safeguarding: It protect the healthcare data that includes all aspects of securing data.

- Vulnerability Management: Vulnerability management tools are required that provides the ability to identify the vulnerabilities in the system.

- Access Control: It follows the principle by "allowing access when in doubt" so that healthcare data can be provided even in the emergency situation.

- Monitoring: Continuous monitoring and quick response would guarantee greater security of the healthcare data. It involves awareness of any data flow and the risk associated with securing big data problem.

Figure 3.6 Challenges that exist in big data healthcare.

3.4 Existing Challenges and Benefits that are Prevalent in Big Data Healthcare

Data management methods involve in big data analysis are constantly developed to capture the data from multiple sources. This section discusses about the challenges that exist in health care data shown in Figure 3.6.

A. Challenges related to Data and Process

They are further divided into following section which are discussed below:

1. **Data Acquisition:** Healthcare data is highly segmented which is derived from multiple heterogeneous sources. An efficient synchronization mechanism is required failing which gaps will be created resulting in misleading of information. Extracting the data from multiple sources and converting it into one global format is a big challenges in healthcare data analytics [41].

2. **Data Storage and Linkage:** Handling health care data is difficult due to its enormous size. It creates the problem of redundancy and it further becomes difficult to analyse the data which is incomplete in nature. Thus, data analysing from multiple sources creates a big problem.

3. **Interoperability:** Since the healthcare data is generated from various devices it supports different formats. To effectively understand the data the devices should communicate in a common data format.

4. **Quality of Data:** Patient data is collected from X-rays, pathology reports, diagnostic images and bedside monitors. There is a need of process to analyse the data which can help in pulling out such type of information .Perhaps, technical issues and human errors may create a dicey data which have a negative impact on patients.

5. **Security and Privacy of Big Data:** Precautionary measures should be adopted to protect patient data .As its misuse can create a problem. Hence, security and privacy of data is another major challenge that need to be taken care off.

B. **Challenges related to Manpower**

The challenges related to Manpower are further divided into following parts;

1. **Talent Deficit:** Means that there is a deficiency of talented or skilled people in big data technologies. Healthcare-specific data scientists and analysts are in high demand for leveraging technologies and machines. But there is a lack of people having the skill and experience in this field.

2. **Retaining Big Data Talent:** It is still challenging and expensive to attract qualified big data expertise with experience in healthcare since, as we all know, our society is lacking in skill. The retention of highly skilled data analysts is also another significant issue that we face.

3. **Human Liaison:** All stages of big data analytics require human involvement in addition to the usage of machines. To offer meaningful answers, experts from many fields must work together using big data techniques.

C. **Challenges related to Domain**

The challenges related to Domain are further divided into following parts:

1. **Temporality:** Time is a key component for taking important healthcare decisions. However, creating algorithms and systems that can use temporal health data is a challenging and uncomfortable undertaking.

2. **Domain Complexity:** The multiplicity of diseases, interaction between patients and providers, difficulty in accurate diagnosis are all come under the complexity of healthcare. The complexity is also affected by the environmental and organizational factors [22].

3. **Interpretability:** The decisions in healthcare plays a vital role since the lives are at stake. The results and logic after the analysis are crucial and plays a crucial role in conclusive healthcare professionals.

D. Challenges related to Managerial/Organizational

The challenges related to Managerial/Organizational are further divided into following sections;

1. **Unclear Purpose:** Uncertain purposes are the first obstacle to implementing big data analytics' promise. Organizations are using big data technologies as a source of competitive benefits. Another difficulty is defining the company goals for using big data analytics.

2. **Technology Disparity:** The majority of companies continue to rely on or use traditional technologies despite the increased interest in digitising the healthcare sector. Nevertheless, because paper-based records may hold so much data, digitising them would take a lot of work, time, and resources. The challenge for humans is this technological disparity.

3. **Identifying Relevant Data and Appropriate Tools:** We must develop methods and tools for collecting and locating pertinent data. It can be difficult for enterprises to locate the right tools and solutions once they are aware of the business.

4. **Timely Actionable Insights:** Using analytics tools and solutions, big data analytics in the healthcare industry produces additional information. Due to the activities they suggest rather than just giving straightforward solutions, actionable insights are more beneficial.

5. **Organizational Resistance:** Another impediment is the slow adoption of technology and organisational hurdles. Due to a lack of awareness of the advantages of adopting big data analytics for both business and human resources, there is little drive to implement big data technology.

6. **Cost Management:** The most crucial concern following the adoption of this extensive technology, tools, and approach is managing the expense of the database system and the infrastructure needed to

store massive amounts of information [29]. The analysis of the data requires a massive computer resources. Large corporations are wary about making such a significant expenditure.

7. **Sharing:** Organizations must share this data with other organisations in order to get the most from big data technology for the community. Big data analytics for health may find difficulties due to data sharing.

3.5 Process Flow of Big Data Analytics in Healthcare

The big data analytics healthcare process flow comprises of following steps as shown in Figure 3.7

- Acquisition of Data

- Storage of Data

- Analytics of Data

Acquisition of Data: The healthcare big data comprises of data in different format such as structured, semi-structured or unstructured. These data are extracted from multiple primary sources such as electronic health records and secondary sources include insurance companies and government sources.

Storage of Data: The storage holds an integral role in big data. With the increase in size of healthcare data an efficient storage mechanism is required. Nowadays, Cloud computing serve as an important tool for storing huge collection of data. The advantages offered by cloud computing are that it removes the need to store enormous data sets. Different cloud platforms that are available are Smart cloud, Azure S3, NoSQL.

Analytics of Data: Data analytics is the process of transforming the raw data into useful information as shown in Figure 3.8. The major big data analytics used in healthcare data are Apache, Hadoop etc.

3.6 Tools and Techniques in Healthcare Data Analytics

The healthcare system is brimming with intelligence, but it needs to change in order to learn more. The most efficient tools and methods are designed to unearth hidden patterns and linkages. According to Kavakiotis et al. [39], a thorough systematic review is performed on the use of machine learning and data mining techniques to perform analytics on diabetes (DM) for prediction and diagnosis, finding complications connected to it, and its linkage

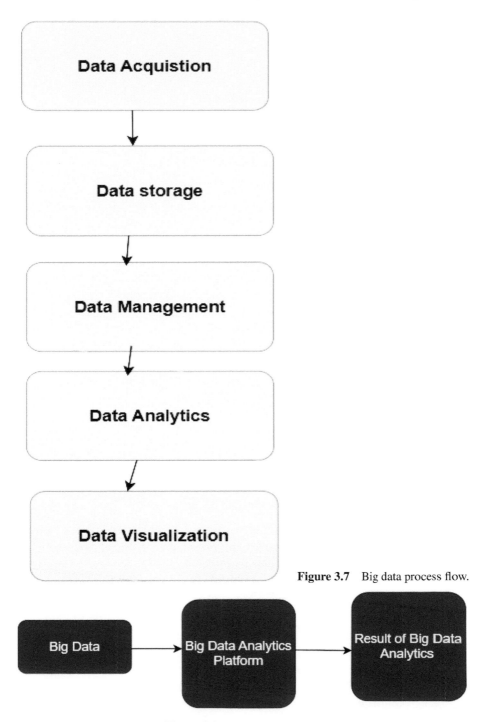

Figure 3.7 Big data process flow.

Figure 3.8 Big data analytics.

Table 3.2   The analytical tools used for analyzing medical data sources.

Cite/tool	HDFS	MapReduce	Hive	Pig	STROM	Intellicius	Hunk
[21]	EHR is input	For analyzing patients data	Analysis	Analysis	Live streaming	Reporting	Reporting

with the genetic background and surroundings will help improve healthcare management.

The Cloud Security Alliance (CSA)-led big data group creates the safe architecture for big data analytics [40]. (Illustrated in Table 3.2). In the healthcare scenario [41], it would be used as the input dataset by the general-purpose analytics platform, which has HDFS with FLUME and SQOOP at the data layer. By looking for similar patterns, Map Reduce and HIVE analyse the data using machine learning techniques. H-Base is used to store multi-structured data. AWS Lambda functions are utilised by STROM, a live streaming service, to promptly notify all affected medical workers of any emergency.

The benefits that are incurred by using Data mining techniques in healthcare sector are:

- It helps to mitigates fraud.

- Increased diagnosis accuracy.

- Better customer interaction.

- Improving Patient Outcome and Safety precautions

With the aid of qualified resources who are familiar with medical terminology and are also used to extract important data from various sources, healthcare organisations may implement Knowledge Discovery in Databases (KDD) [42] using frameworks provided in [43]. Through appropriate data mining and machine learning procedures where particular algorithms are utilised to extract patterns from data to further clarify, KDD is able to identify a meaningful pattern in the data to generate crucial solutions.

The tasks of class definition, classification, relationships, prediction, clustering, and time series analysis are completed through data mining. One method for processing data in a multidimensional capacity is online analytical processing (OLAP) [44].

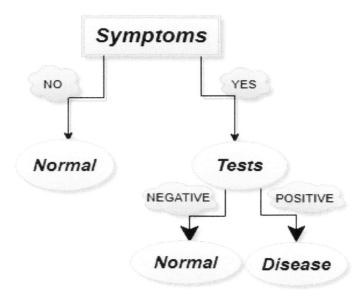

Figure 3.9 Classification rules extracted from a decision tree for diagnosis.

3.7 Classification and Clustering Techniques

- *Rule-set classifiers*
  'If-then' rules are used to extract information. IF *condition* THEN *conclusion that* means IF part consist of questions and THEN part gives the result. [42] E.g. if a symptom is discovered positive then diagnosis of the disease takes place.

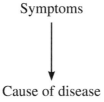

- *Decision tree*
  Information is represented as a network of nodes and branches. Figure 3.9 illustrates a decision tree, which is a flowchart-like tree structure where each internal node indicates a test on an attribute, each branch represents a test result, and each leaf node (terminal node) stores a class label.

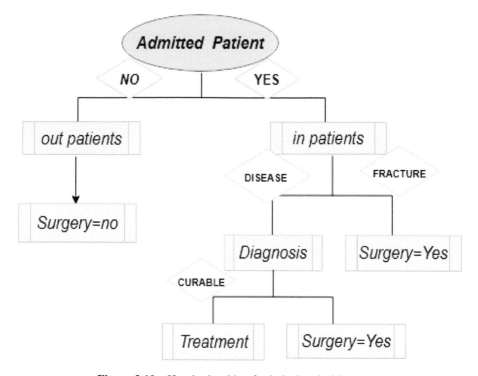

Figure 3.10   Hunt's algorithm for inducing decision trees.

The decision tree algorithms C4.5, ID3, HUNTS, CART, SLIQ, SPRINT, and others are the most used ones [13]. Figure 3.10 depicts the Hunt technique for generating decision trees.

Neural networks algorithms

- RNNs (Recurrent Neural Networks)
  A recurrent neural network (RNN) is a type of artificial neural network in which connections between nodes can form a cycle, allowing the output of some nodes to influence the input received by other nodes in the same network. The method forecasts using the patient's entire medical history and is tested using real EHR data from 250k patients over the last 8 years as shown in Figure 3.11 [45].

- *Neuro-fuzzy*
  To build the fuzzy neural network, the stochastic back propagation technique [46] is applied. First, we assign random weights to the links.

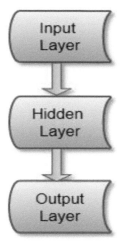

Figure 3.11 A simple RNN diagram.

The input, output, and error rate are then calculated. Thirdly, each node determines its own certainty measure (c), which is used to calculate uncertainty.

- *Artificial Neural Networks (ANN)*
  They are composed of several neurons, and neurons are the fundamental building block of a perceptron. These neurons are linked by connections that control the transmission of information between them. Artificial neurones two different sorts of signals [47]:

  (i) excitatory

  (ii) inhibitory to activate

  The most likely signal to be conveyed will be an excitatory if the input is excitatory. Most likely, an inhibitory signal would propagate if the input signal is one. Figure 3.12 displays the neuron's fundamental model.

3.8 Limitations and Strength

Due to its density, big data analysis itself causes congestion. Because of this, medical professionals are looking to computer scientists for information and solutions so they can learn from data. In the discipline of data science, there are a few emerging technologies, including Hadoop, unsupervised learning, graph analytics, and natural language processing. The research community's

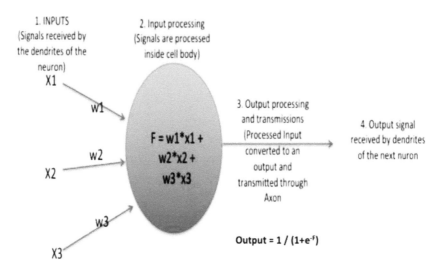

Figure 3.12   Basic model of a single neuron [47].

intuitive skills have been consumed by the large, distinct data in the medical profession. Through the critical use of algorithms, correlations are created with all associative elements and attributes. Thus, there is a great opportunity to promote medicine as an information science and establish the groundwork for a learning healthcare system. Accessing layer, Business implementation layer, Support layer are the four levels that make up the Health information service model as shown in Figure 3.13.

Clinical staff generally lacks a clear understanding of the standardised and exact picture of information on patients, and benefits and hazards are frequently unclear. Even if there is data to support a given conclusion in a specific context, it may not necessarily be relevant to the patient.

The algorithms and models used in big data and conventional health-care systems are very similar. Regarding user interface, there is a significant difference between standard analytics tools and those used for big data analytics as shown in Figure 3.14. The user-friendliness of big data analytics tools has not kept up with that of traditional technologies. They are complicated, involve a lot of programming, and call for a wide range of abilities. Tools for big data analytics are open source, making them less complicated.

It is not at all easy to comply with the needs of stakeholder in healthcare whether they meet the needs of healthcare stakeholders, including patients,

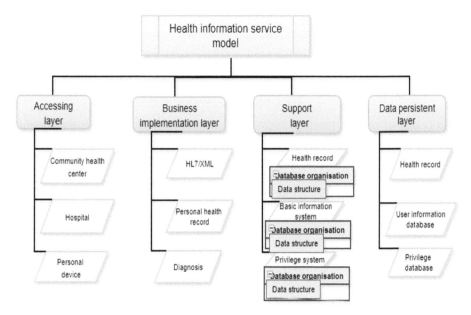

Figure 3.13 Functional framework of a health information service model.

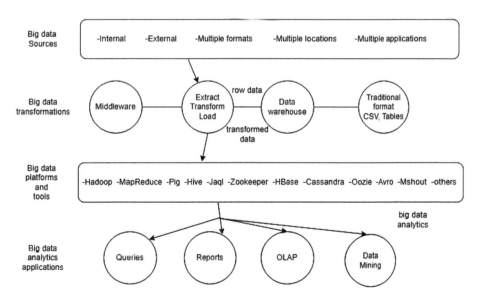

Figure 3.14 Conceptual architecture of big data analytics.

medical professionals, administrators, and other regulatory bodies, is by no means simple. Furthermore, it is currently too late to adequately educate the medical community for the aim of precision medicine [48-52], which will be developed for personalised healthcare using personalised information through learning from physicians and practitioners' input. Our strength stems from our drive to make the learning healthcare system (LHS) a reality. The usage of open source platforms is now highly encouraged, and healthcare data analytics using Hadoop/Map Reduce can be done on the cloud. With the help of internet technology healthcare data is streamed to store large data sets. Due to the availability of large number of cloud storage platforms namely Azure S3, NoSQL and Smart cloud data can easily be store and retrieved. [20]

3.9 Conclusion

Healthcare is an important sector which has shown how three V related to data namely variety, velocity and volume are mapped. The recent advancement in information technology has led to the ability of collecting enormous data in healthcare sector. Healthcare data are collected from wide range of sources such as images, sensors, biomedical literature and electronic records. The heterogeneity of data collection and its representation causes several challenges in performing pre-processing and analysis of the data in different perspective. The vital elements which are required for performing healthcare data analytics are Data Elements, Functional Elements, Human element and Security Element. In this paper we emphasises on systematic review of identifying the strength and limitation that exist in big data analytics in Health care sector. We further, discussed the different tools and techniques that are used in big data analytics. The basic objective of the chapter is to provide the guiding opportunities for successful implementation of big data technology in different sectors of healthcare.

References

[1] A. McAfee, E. Brynjolfsson, T. H. Davenport, D. J. Patil, and D. Barton, "Big data: the management revolution," Harvard Business Review, vol. 90, no. 10, pp. 60–68, 2012.
[2] C. Lynch, "Big data: how do your data grow?" Nature, vol. 455, no. 7209, pp. 28–29, 2008.

[3] A. Jacobs, "The pathologies of big data," Communications of the ACM, vol. 52, no. 8, pp. 36–44, 2009.

[4] P. Zikopoulos, C. Eaton, D. deRoos, T. Deutsch, and G. Lapis, Understanding Big Data: Analytics for Enterprise Class Hadoop and Streaming Data, McGraw-Hill Osborne Media, 2011.

[5] J. Manyika, M. Chui, B. Brown, J. Bughin, R. Dobbs, C. Roxburgh, and A. H. Byers. Big data: The next frontier for innovation, competition, and productivity. McKinsey Global Institute Report, May 2011.

[6] L. A. Celi, R. G. Mark, D. J. Stone, and R. A. Montgomery, "'Big data' in the intensive care unit: closing the data loop," American Journal of Respiratory and Critical Care Medicine, vol. 187, no. 11, pp. 1157–1160, 2013.

[7] L. Hood and N. D. Price, "Demystifying disease, democratizing health care," Science Translational Medicine, vol. 6, no. 225, Article ID 225ed5, 2014.

[8] Castro EM, Van Regenmortel T, Vanhaecht K, Sermeus W, Van Hecke A. Patient empowerment, patient participation and patient-centeredness in hospital care: a concept analysis based on a literature review. Patient Educ Couns. 2016;99(12):1923–39.

[9] Marconi K, Dobra M, Thompson C. The use of big data in healthcare. In: Liebowitz J, editor. Big data and business analytics. Boca Raton: CRC Press; 2012. p. 229–48.

[10] Catherine M. DesRoches et al. Electronic health records in ambulatory carea national survey of physicians. New England Journal of Medicine 359(1):50–60, 2008.

[11] G. Acampora, D. J. Cook, P. Rashidi, A. V. Vasilakos. A survey on ambient intelligence in healthcare,Proceedings of the IEEE, 101(12): 2470–2494, Dec. 2013.

[12] Robert F. Woolson and William R. Clarke. Statistical Methods for the Analysis of Biomedical Data, Volume 371. John Wiley & Sons, 2011.

[13] S. M. Meystre, G. K. Savova, K. C. Kipper-Schuler, and J. F. Hurdle. Extracting information from textual documents in the electronic health record: A review of recent research. Yearbook of Medical Informatics, pages 128–144, 2008.

[14] L. Jensen, J. Saric, and P. Bork. Literature mining for the biologist: From information retrieval to biological discovery. Nature Reviews Genetics, 7(2):119–129, 2006.

[15] Zhang, R., G. Simon, and F. Yu. 2017. "Advancing Alzheimer's research: a review of Big data promises." International Journal of Medical Informatics 106: 48–56. doi:10.1016/j. ijmedinf.2017.07.002.

[16] Bates, D. W., Saria, S., Ohno-Machado, L., Shah, A., & Escobar, G. (2014). Big data in health care: Using analytics to identify and manage high-risk and high-cost patients. Health Affairs, 33(7), 1123–1131. doi:10.1377/ hlthaff.2014.0041

[17] O'Driscoll, A., Daugelaite, J., & Sleator, R. D. (2013). 'Big data', Hadoop and cloud computing in genomics. Journal of Biomedical Informatics, 46(5), 774–781. doi: 10.1016/j.jbi.2013.07.001

[18] Quinn, C. C., Clough, S. S., Minor, J. M., Lender, D., Okafor, M. C., & Gruber-Baldini, A. (2008). WellDocTM mobile diabetes management randomized controlled trial: Change in clinical and behavioral outcomes and patient and physician satisfaction. Diabetes Technology & Therapeutics, 10(3), 160–168. doi:10. 1089/dia.2008.0283.

[19] Wamba, S. F., Anand, A., & Carter, L. (2013). A literature review of RFID-enabled healthcare applications and issues. International Journal of Information Management, 33(5), 875–891. doi:10.1016/j. ijinfomgt.2013.07.005.

[20] Andreu-Perez, J., Poon, C. C., Merrifield, R. D., Wong, S. T., & Yang, G. Z. (2015). Big data for health. IEEE Journal of Biomedical and Health Informatics, 19(4), 1193–1208. doi:10.1109/JBHI.2015.2450362.

[21] Hilario, M., Kalousis, A., Pellegrini, C., & Muller, M. € (2006). Processing and classification of protein mass spectra. Mass Spectrometry Reviews, 25(3), 409–449. doi:10.1002/mas.20072.

[22] Murdoch, T. B., & Detsky, A. S. (2013). The inevitable application of big data to health care. JAMA, 309(13), 1351–1352. doi:10.1001/ jama.2013.393.

[23] Barrett, M. A., Humblet, O., Hiatt, R. A., & Adler, N. E. (2013). Big data and disease prevention: from quantified self to quantified communities. Big data, 1(3), 168–175.

[24] Gligorijevic, V., Malod-Dognin, N., & Przulj, N. (2016). Integrative methods for analyzing big data in precision medicine. Proteomics, 16(5), 741–758. doi:10.1002/ pmic.201500396

[25] Kuiler, E. W. (2014). From big data to knowledge: An ontological approach to big data analytics. Review of Policy Research, 31(4), 311–318. doi:10.1111/ropr.12077

[26] Bardhan, I., Oh, J. H., Zheng, Z., & Kirksey, K. (2015). Predictive analytics for readmission of patients with congestive heart failure. Information Systems Research, 26(1), 19–39. doi:10.1287/isre.2014.0553.

[27] Basole, R. C., Braunstein, M. L., Kumar, V., Park, H., Kahng, M., Chau, D. H. (P.)., Thompson, M. (2015). Understanding variations in pediatric asthma care processes in the emergency department using visual analytics. Journal of the American Medical Informatics Association, 22(2), 318–323.

[28] Ajorlou, S., Shams, I., & Yang, K. (2015). An analytics approach to designing patient centered medical homes. Health Care Management Science, 18(1), 3–18. doi:10. 1007/s10729-014-9287-x

[29] Delen, D. (2009). Analysis of cancer data: A data mining approach. Expert Systems, 26(1), 100–112. doi:10.1111/j. 1468-0394.2008.00480.x

[30] Liu, P., & Wu, S. (2016). An agentbased simulation model to study accountable care organizations. Health care management science, 19(1), 89–101.

[31] Holzinger, A., & Jurisica, I. (2014). Knowledge discovery and data mining in biomedical informatics: The future is in integrative, interactive machine learning solutions. In Interactive knowledge discovery and data mining.

[32] Althebyan, Q., Yaseen, Q., Jararweh, Y., & Al-Ayyoub, M. (2016). Cloud support for large scale e-healthcare systems. Annals of Telecommunications, 71(9–10), 503–515. doi:10.1007/s12243-016-0496-9.

[33] Rastogi, M., Vijarania, M., & Goel, N. (2023, February). Implementation of Machine Learning Techniques in Breast Cancer Detection. In International Conference On Innovative Computing And Communication (pp. 111-121). Singapore: Springer Nature Singapore.

[34] Abbas, A., Ali, M., Khan, M. U. S., & Khan, S. U. (2016). Personalized healthcare cloud services for disease risk assessment and wellness management using social media. Pervasive and Mobile Computing, 28, 81–99. doi:10.1016/j.pmcj.2015.10.014.

[35] Boulos, M. N. K., Sanfilippo, A. P., Corley, C. D., & Wheeler, S. (2010). Social Web mining and exploitation for serious applications: Technosocial Predictive Analytics and related technologies for public health, environmental and national security surveillance. Computer Methods and Programs in Biomedicine, 100(1), 16–23. doi:10.1016/j.cmpb.2010.02.007.

[36] Groves, Peter; Kayyali, Basel; Knott, David; Kuiken, Steve Van The 'big data' revolution in healthcare: Accelerating value and innovation, 2016.

[37] Rastogi, M., Vijarania, D. M., & Goel, D. N. (2022). Role of Machine Learning in Healthcare Sector. Available at SSRN 4195384.

[38] Grover, M.: Processing frameworks for Hadoop. O'Reilly Media (2015). https://www.oreilly.   com/ideas/processing-frameworks-for-hadoop. Accessed 14 Mar 2019.

[39] Vijarania, M., Udbhav, M., Gupta, S., Kumar, R., & Agarwal, A. (2023). Global Cost of Living in Different Geographical Areas Using the Concept of NLP. In Handbook of Research on Applications of AI, Digital Twin, and Internet of Things for Sustainable Development (pp. 419-436). IGI Global.

[40] Gupta, S., Vijarania, M., & Udbhav, M. (2023). A Machine Learning Approach for Predicting Price of Used Cars and Power Demand Forecasting to Conserve Non-renewable Energy Sources. In Renewable Energy Optimization, Planning and Control: Proceedings of ICRTE 2022 (pp. 301–310). Singapore: Springer Nature Singapore.

[41] Archenaa J, Anita EM (2015) A survey of big data analytics in healthcare and government. Procedia Comput Sci 50:408–413

[42] Wu X et al (2008) Top 10 algorithms in data mining. Knowl Inf Syst 14(1):1–37

[43] Zhao J, Papapetrou P, Asker L, Boström H (2017) Learning from heterogeneous temporal data in electronic health records. J Biomed Inform 65:105–119.

[44] Demirkan H, Delen D (2013) Leveraging the capabilities of service-oriented decision support systems: putting analytics and big data in cloud. Decision Support Systems, vol 55, no 1, pp 412–421.

[45] Miah SJ, Hasan J, Gammack JG (2017) On-cloud healthcare clinic: an e-health consultancy approach for remote communities in a developing country. Telemat Inform 34(1):311–322.

[46] Srinivas K, Rani BK, Govrdhan A (2010) Applications of data mining techniques in healthcare and prediction of heart attacks. Int J Comput Sci Eng 2(02):250–255.

[47] V.K.Harikrishnan,MeenuVijarania, AshimaGambhir, Diabetic retinopathy identification using autoML, Computational Intelligence and Its Applications in Healthcare, 2020, Pages 175–188

[48] Krumholz HM (2014) Big data and new knowledge in medicine: the thinking, training, and tools needed for a learning health system. Health Aff (Millwood) 33(7):1163–1170.

[49] Gupta, S., & Vijarania, M. (2013). Analysis for Deadlock Detection and Resolution Techniques in Distributed Database. International Journal of Advanced Research in Computer Science and Software Engineering, 3(7).

[50] Srinivas K, Rani BK, Govrdhan A (2010) Applications of data mining techniques in healthcare and prediction of heart attacks. Int J Comput Sci Eng 2(02):250–255.

[51] M Vijarania, N Dahiya, S Dalal, V Jaglan, WSN Based Efficient Multi-Metric Routing for IoT Networks, Green Internet of Things for Smart Cities, 2021 , Edition1st Edition, First Published2021,CRC Press, 249–262.

[52] Nanda, A., Gupta, S., & Vijrania, M. (2019, June). A comprehensive survey of OLAP: recent trends. In 2019 3rd International Conference on Electronics, Communication and Aerospace Technology (ICECA) (pp. 425–430). IEEE.

4

Role of Big Data Analytics in the Cloud Applications

V. Sakthivel[1*], Sriya Nanduri[2], and P. Prakash[2]

[1]KADA, Konkuk University, Seoul, South Korea; Vellore Institute of
Technology, India
[2]Vellore Institute of Technology, India

Email: mvsakthi@gmail.com; shriyananduri@gmail.com;
prakash.p@vit.ac.in

Abstract

One of the most popular technologies which are relevant in today's IT industry would be cloud computing and big data analytics. Data are extremely
large and can be used to analyze underlying trends, patterns, and correlations,
which can be of business value is called big data. The datasets are so large
that their volumes increase exponentially over time. The process of analyzing
patterns in big data using advanced analytical techniques is called big data
analytics. Whereas, cloud computing is the provision of computational services on demand, over the Internet. These services and the data stored are run
on remote servers present in data centers across the world. These two concepts
may seem to be very different from each other, but we can combine these two
technologies to provide various services to people all over the Internet. Cloud
Computing is now enabling the big data industry to achieve what it has not been
able to in the past. To achieve the goals of big data analytics, cloud computing
can be used as a platform. In cloud computing, the data which is remotely
stored is processed in real-time, interpreted, and delivered appropriately to
users. This improves the quality of customer services companies provide to
their customers. Cloud can take in huge amounts of data and analyze it in a
fraction of seconds. Moreover, since the type of data used in big data analytics
is not always organized or in a standard format, the artificial intelligence and

69

machine learning technologies of cloud computing can be used to convert the data into a standard format. This can prove to be very effective and economical for business enterprises. It is because traditional on-premise data centers are harder to manage and need a lot of maintenance by companies. Therefore, as an organization grows, it is better to shift its business model toward cloud. With these two technologies combined, companies can achieve economies of scale with minimal investment and improve overall customer experience.

4.1 Introduction

4.1.1 About cloud computing

Cloud computing is the on-demand provision and availability of IT services and resources such as computing power, storage, databases, analytics, intelligence, software, and networking via the Internet to achieve economies of scale because of pay-as-you-go pricing models. This will help you lower your operational costs and manage your business, and infrastructure with ease and more efficiently. The companies that offer these services are called cloud service providers (CSPs).

4.1.2 About big data analytics

Big data analytics is the process in which you find patterns, behaviors, trends, and correlations in massive datasets which provide business value using advanced analytical methods that cannot be achieved with traditional data management techniques and tools.

4.2 Cloud Computing

A cloud can be deployed in various ways (Figure 4.1). Some of the deployment models of the cloud are [3]:

1. **Private cloud**: This is installed on an organization's property and solely used inside the premises. Through this model, they get dedicated infrastructure, which is not shared by any other organization or individual. It is usually the first stepping stone for companies who wish to transition to the cloud. For security, network bandwidth is not an issue for this type of deployment. The cost is borne by the organization and the CSPs do not provide any management services.

2. **Public cloud**: In this type of cloud deployment, the infrastructure is established in the premises of the CSP, an organization or an individual

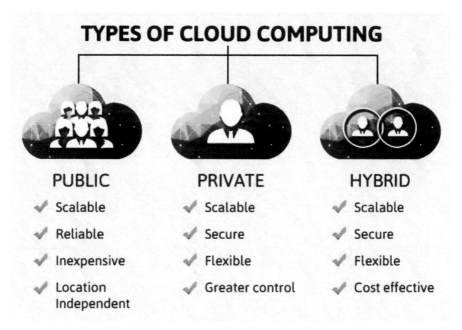

Figure 4.1 Types of cloud computing [23].

can rent cloud services from CSPs on-demand and on a pay-as-you-go pricing model which makes it the most economical of all the types of cloud deployment models. Multiple users share the resources provided by the CSPs via the Internet. You have to pay only for the services and resources you have used.

3. **Hybrid cloud**: It comprises both the characteristics of private and public cloud. This type of model can be used by organizations when they require scalability of their services during business need fluctuations so that they can connect to the public cloud at any time to provide services to their customers in times of high demand. This model allows sharing of data between private and public clouds. It also allows them to safeguard their sensitive data on the private cloud while using the public cloud to offer services.

There are also three main types of computing services (Figure 4.2) [2]:

1. **Software as a service**: SaaS is a service that delivers a software program via the Internet to customers using a multitenant architecture, which is managed by the CSP. The customer does not pay any upfront costs to use the service provided by CSP (like software licensing and

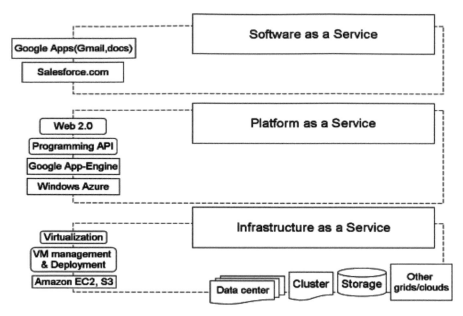

Figure 4.2   Cloud service stack [1].

costs for the servers). The company which provides the software as a service also does not have to invest a lot compared to conventional hosting methods and the maintenance is much easier this way. In this kind of service, the service provider manages and executes all required hardware and software. For instance, Dropbox is a popular tool for sharing files and collaboration that offers cloud storage. This online tool can be used for backup purposes. The operating systems it is available on are macOS, Linux, and Windows. It provides a free storage facility of 2GB under the free tier, and over 100GB storage under paid plans. 350 GB is offered by Dropbox for Teams for the users. The service uses Amazon S3 for storage incorporated with AES and SSL for security.

2.   **Platform as a service**: PaaS is a service that provides a development environment to the customers. The hardware and software platforms are managed by the CSP but the user has to manage the applications running on the platform. The software platform includes a variety of tools for business intelligence, DBMS (database management systems), and other middleware. This service can support scalability, availability, transactions, and authentication. The entirety of the application lifecycle is managed by PaaS which includes planning, development, testing, deployment, and maintenance. An example of PaaS could be

AWS elastic beanstalk which helps you manage your web application by reducing its complexity.

3. **Infrastructure as a Service**: IaaS is a kind of service with the most freedom that a CSP can provide you with. It gives even OS-level control to the developers. The CSPs will manage the infrastructure which is the hardware component (servers, networking, virtualization, and storage), and the user can the infrastructure via an API. The user has to manage the operating system updates and middleware. For example, Amazon EC2 and S3 are some of the IaaS services provided by AWS.

Cloud computing is widely different from traditional computing methods. Following are the properties of cloud computing [4]:

1. **Resource pooling**: The resources of the CSPs are combined and shared among numerous clients with different services to each of them according to the customer's requirements. The physical and virtual resources are reassigned and allocated depending on the demand for storage, bandwidth, processing power, etc. This process is almost instant and the client does not face any issues with the overall experience. The client does not have any control over the location of the resources.

2. **Ad hoc self-service**: The user can continuously track the capability, computing functionalities, server uptimes, and allotted network storage constantly. The user can demand allotment of additional resources almost instantly with no or minimum upfront.

3. **Easy maintenance**: Through cloud computing, management becomes easier, and since there is almost no downtime except in a few cases, where there is also a relatively smaller downtime than traditional methods of hosting and providing the same services. The servers get frequent software updates and patch fixes for enhancing their potential and better performance and experience with each updated version.

4. **Large network access**: With this feature, any client can utilize any device via the Internet to access the cloud, and upload or transfer data to the cloud from anywhere across the world.

5. **Availability and resilience**: Resilience means that the cloud can recover from any disruptions. It is measured by how fast its databases, servers, and networks can restart and recover from any disruption. One goal of the cloud is high availability, which is the provision of services and resources to your customers at any time from anywhere using any device with any Internet connection.

6. **Automatic system**: The cloud can automatically install, configure, and maintain itself. This reduces the dependence on human supervision over the system. However, some supervision is still needed over the resources which help in automation for the smooth running of services.

7. **Economical**: With traditional on-premise data centers, the company has to purchase servers and maintain them by themselves. If the hardware becomes, obsolete, the cost of replacement also has to be borne by them. Therefore, operational costs are relatively higher in traditional on-premise data centers when compared to purchasing cloud services from a CSP. You only have to pay for the services and resources you opt for and use. All this saves the companies with lower monthly or annual costs.

8. **Security cloud**: CSPs usually store multiple copies of your data in various locations, so that if the data gets lost in one location because of any reason, you can use the other backups of your data instead. This ensures zero interruption in your services. The data are also encrypted in the servers and therefore cannot be used by any malicious entity.

9. **Pay-as-you-go**: The users have to only pay for the computational services, network, and storage they are using with no upfront costs. These services are allocated often free of cost which makes it quite economical.

10. **Measured service**: This is one of the services both CSPs and customers use to measure and report what resources are being used and for what purpose. Usage is calculated by charge-per-use. This is used to monitor billing, costs, and usage of resources by the clients. You can also check how much you will be paying by the end of the month, year from using the set of services you are using.

Motivation toward cloud computing in recent times [1]

The recent emergence of cloud computing is a result of the following business and technological trends:

Productivity: Parallelized batch processing can be done on petabytes of data efficiently using the cloud. This reduces time in performing computations. For example, a streaming analytics solution called Google cloud dataflow combines batch and stream data processing with data freshness.

Recovery and security of data: CSPs store multiple backups of your data in different availability zones for disaster recovery. Public cloud services have invested a lot in the security of the cloud which has lifted the concern of security breaches. They provide a setting that is more secure than what an on-premise data center can offer.

Easy access and low maintenance: With the cloud, you do not have to maintain the infrastructure. All the resources are accessed virtually. Users have access to their data by using devices from anywhere in the world.

Need for compute power: A lot of desktop applications such as Matlab, Anaconda, etc., and building machine learning models requires a lot of computing power and therefore might not be suitable to operate on a desktop computer for medium to large-scale projects. This problem can be solved through cloud computing as it provides the necessary computation services to achieve the results. For example, Google Colab which is based on Jupyter Notebook allows users to develop Python-based machine learning models and also collaborate with other peers to edit the notebooks.

Examples for cloud computing platforms
Simple storage service: Amazon S3 is a storage service which is object-based, which means that if you wish to make a change in the stored file, you have to re-upload the entire file after the change has been made. Data are stored in objects in containers called buckets. This service provides 11 9s of durability. Each object can be of a maximum size of 5 TB. S3 buckets can be used for storing snapshots, archiving, server logs, images, videos, documents, etc. There is a restriction of 100 buckets per account. The data can be accessed through APIs, SDKs, and AWS management console with fine-grained access through AWS IAM (identity and access management).

Elastic compute cloud: Amazon EC2 (elastic cloud compute) is a service which provides infrastructure (virtual machines) to the users. This means you do not have to pay for the hardware costs for your application and only pay for the computing power you have used. It allows you to choose the OS of your choice, security and network settings, and the size and storage options for the EC2 instance you want to launch. You can host your applications on these instances. You can decrease or increase the number of EC2 instances you launch depending on the demand (Figure 4.3).

Big data analytics
Big data analytics is the process in which you find patterns, behaviors, trends, and correlations in massive datasets which provide business value

Figure 4.3   Amazon EC2 and Amazon S3 [21, 22].

using advanced analytical methods that cannot be achieved with traditional data management techniques and tools.

To understand the idea behind big data analytics, we have to compare it with regular data analytics.

Traditional data analytics: This type of analysis usually is done in offline mode after a certain period of time or a specific event. Only structured data which has been maintained in a centralized database is analyzed. Such type of data are only generated at the enterprise level therefore the volume is small. Since it is structured data, it can be easily manipulated by simple functions. SQL (Structured Query Language) is used to access and manage the data. Examples: Financial data, organizational data, web transaction data, etc.

Big data: The analysis for big data happens in real-time and discoveries from the analysis can be deduced almost instantaneously. For such analysis, very large and complex databases are used which are very difficult to process using primitive tools. It deals with all kinds of data, be it structured, unstructured, or semi-structured. Meaningful data are extracted from large amounts of unstructured data for analysis.

The four main categories of big data analytics are listed below.

- One popular type of analytics that enables you to learn what occurred and when is **descriptive analytics**.

- **Diagnostic analytics** locates patterns and linkages in the available data to explain why and how something occurred.

- **Predictive analytics** analyses historical data to identify trends and forecast likely future events.

- **Prescriptive analytics** provide detailed advice on what should be improved.

Four main dimensions correspond to the main features of big data. The following definitions apply to these dimensions:

1. **Volume**: There is only one size of big data: large. Enterprises can easily accumulate terabytes or even petabytes of data, which is a data avalanche.

2. **Variety**: Big data encompasses all forms of unstructured data, including click streams, log files, text, audio, and video, in addition to structured data.

3. **Veracity**: The vast volumes of data gathered for big data applications may result in statistical errors and incorrect interpretations of the data. Value depends on the information's accuracy.

4. **Velocity**: It is the measure of how fast the data is being generated, processed, stored, managed, and provided to meet the demands.

These big data 4Vs outline the route to analytics, and each has inherent value in the quest to find value.

Big data analysis techniques [7]

1. Association rule learning

Unsupervised learning techniques like association rule learning examine if one data element depends on another and build it properly to be more cost-effective. It looks for intriguing relationships or associations between the dataset's variables. To discover intriguing relationships between variables in the database, different guidelines must be followed (Figure 4.4).

The most crucial method of machine learning is association rule learning, which is used in continuous manufacturing, market basket research, web usage mining, etc. Several large retailers employ this method of market basket research to discover the connections between products.

Although data mining is the use of an algorithm to observe patterns on mostly structured datasets in a knowledge discovery process, web mining can be seen as the application of modified data mining methods to the Internet.

Web mining offers the unique ability to support a variety of different data kinds. The web has many features that provide a variety of ways for the mining process, including text-based pages, hyperlinks connecting pages, and the ability to track user behavior through web server logs.

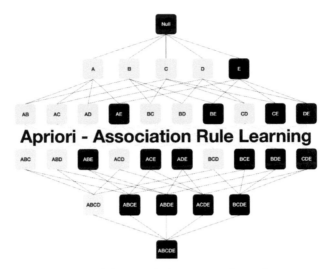

Figure 4.4   Associative rule learning [16].

2. Classification tree analysis

To classify remotely sensed and auxiliary data in support of the mapping and study of land cover, classification tree analysis (CTA) is a sort of machine learning algorithm. The class of any object can be determined by using classification tree analysis, which is a mapping of binary decisions. It is different from a decision tree since it is more of a categorical decision-making tree. Another type of decision tree that produces numerical decisions is the regression tree.

Determining the categories to which a new observation belongs is statistical classification. It needs a training set of accurate observations or historical data (Figure 4.5).

3. Genetic algorithms

Search algorithms called genetic algorithms were developed as a natural extension of Darwin's theory of evolution (Figure 4.6).

- Using the theory of evolution, genetic algorithms can overcome issues with traditional algorithms.

- Simulation of natural selection, mutation, and reproduction, these algorithms can produce high-quality solutions for a variety of issues, including search optimization.

Darwin's theory of evolution states that a population of people that differ from one another is maintained via evolution (variation). Better-adapted

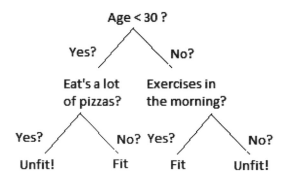

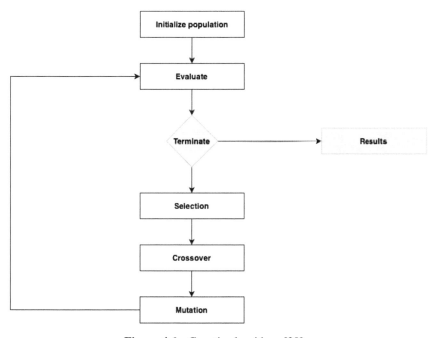

Figure 4.5 Classification tree analysis [17].

Figure 4.6 Genetic algorithms [20].

individuals have a higher likelihood of living, reproducing, and passing on their characteristics to the following generation (survival of the fittest).

4. Machine learning
Machine learning (ML) is a form of artificial intelligence (AI). It can be used for making programs forecast outcomes more precisely by making use of

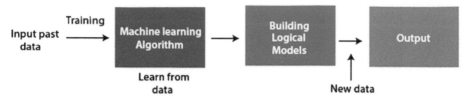

Figure 4.7 Machine learning [19].

datasets. To predict new output values, these algorithms need historical data as input (training data). The models always continuously learn as the application grows thus increasing its accuracy over time.

To build mathematical models and generate predictions based on prior knowledge or data, machine learning employs a variety of methodologies. In addition to speech recognition, spam detection for emails, recommendation systems, and image identification, it is also used for a wide range of other activities.

Machine learning (Figure 4.7) can be classified into:

- **Supervised learning**: In this branch of ML, the algorithm is provided with a labeled training dataset containing both the inputs and outputs while it tries to determine a mechanism to arrive at these results. The algorithm tries to find hidden patterns and makes observations to accurately make predictions. The predictions, if wrong, are corrected by humans, till a satisfactory level of precision is achieved.

- **Unsupervised learning**: In this branch of ML, there is no labeled data fed to the algorithm, rather the algorithm has to find relationships and correlations through analysis. The algorithm tries to arrange and organize the given data by grouping them into clusters to find their structure.

- **Reinforcement learning**: In this branch of ML, the algorithm is given a set of conditions and parameters. It is about choosing an appropriate path to maximize the gains. The algorithm tries to learn from its past mistakes through a trial-and-error method. Unlike supervised learning, the algorithm is not corrected with the right answer but instead is given rewards or punishment accordingly.

5. Regression analysis
Regression analysis is a statistical technique that is used to estimate the correlation between two or more variables. It can be used to simulate strength of

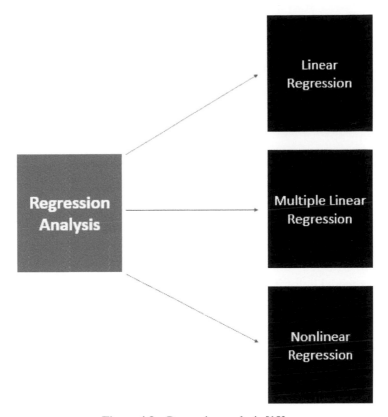

Figure 4.8 Regression analysis [18].

the relation of the variables and predict the future relationship between the variables.

Various types of regression analysis include simple linear, multiple linear, and nonlinear. Simple linear and multiple linear models are the most used of all. For complicated datasets, nonlinear regression analysis is used.

Regression analysis is used extensively in many fields such as finance. It can be used to determine how customer satisfaction might influence customer loyalty (Figure 4.8).

6. Sentiment analysis
The term "opinion mining" also applies to sentiment analysis. It uses natural language processing (NLP) to determine if data is negative, positive, or neutral. Such analysis of textual data is a popular practice among enterprises for monitoring how their products are being perceived by their clients through

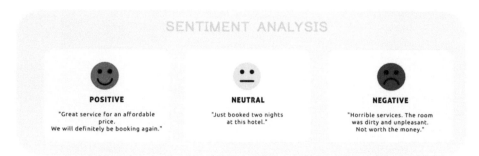

Figure 4.9   Sentiment analysis [15].

online reviews and to understand the requirements of their target audience in a better way (Figure 4.9).

a. **Fine-Grained**

This model helps in polarity precision determination. The polarity categories of extremely positive, positive, neutral, negative, or very negative can all be used in sentiment analysis. Fine-grained sentiment analysis can be used for studying reviews and ratings. For instance: a rating scale of 1 to 10 (from negative to positive).

b. **Aspect-Based**

Aspect-based analysis digs deeper while fine-grained analysis helps you identify the general polarity of your customer reviews. It aids in identifying the specific topics that individuals are discussing.

c. **Emotion detection**

Emotion detection, as its name suggests, aids in emotion recognition. These emotions include rage, grief, joy, and gratification as well as fear, concern, and panic. Lexicons, or collections of words that express particular emotions, are frequently used by emotion detection systems. Robust machine learning (ML) algorithms are also used by some sophisticated classifiers.

7. Intent analysis

Enterprises can save a lot of resources by accurately identifying consumer intent. Most of the time businesses pursue clients who do not have immediate plans to make a purchase. This can be resolved through intent analysis.

The intent analysis aids in determining whether a buyer is intending to make a purchase or is merely looking around.

Variability of Centrality Measures

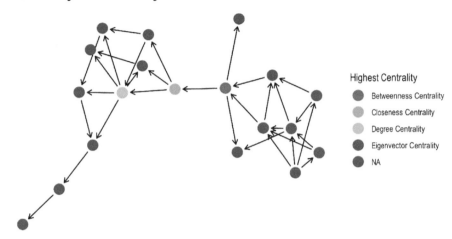

Highest Centrality
- Betweenness Centrality
- Closeness Centrality
- Degree Centrality
- Eigenvector Centrality
- NA

Figure 4.10   Social network analysis [14].

You may follow a customer and advertise to them if they are prepared to make a purchase. You can save resources by avoiding advertising to customers who do not wish to make a purchase.

8. Social network analysis

To analyze a social network both quantitatively and qualitatively, social network analysis can be used (SNA). Graph theory is extensively used to achieve this. Each network is analyzed on the basis of its nodes and its links or edges that connect them (Figure 4.10).

The Internet can be considered as a vast network. There are several webpages linked to other pages. Therefore, it can be used for search engine optimization by analyzing the relationship between such pages.

Some of the indices used on a social network are degree centrality, betweenness centrality, closeness centrality, and eigenvector.

Big data analytics processes [10]

See Figure 4.11.

1. **Data ingestion**

   From locating the sources to obtaining big data, the method differs from business to business. To ensure prompt processing, data collection frequently takes place in real-time or very close to real-time. With today's technologies, it is possible to collect both structured (data that

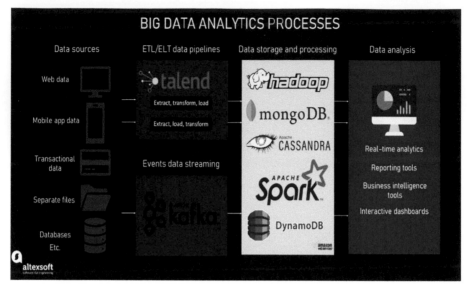

Figure 4.11   Big data analytics processes [10].

is typically in tabular formats) and unstructured (all types of data formats) data from a variety of sources, such as websites, mobile apps, databases, customer relationship management systems (CRMs), IoT devices, and more.

ETL or ELT data pipelines are designed to transfer data from sources to centralized repositories for additional storage and processing. Raw data must go through the extraction, transformation, and loading (ETL) stages. ELT enables data transformation after it has been fed into a target system, as opposed to the ETL strategy, which occurs before data reaches a target repository like a data warehouse.

2.   **Data storage and processing**
Data can be migrated to storage like cloud data warehouses or data lakes, depending on how complicated the data is, so that business intelligence tools can access it as needed. Numerous contemporary cloud-based solutions typically contain client infrastructure, computation, and storage components. Data from various sources can be partitioned by storage layers for further optimization and compression. The groups of processing engines known as compute layers are utilized to carry out any computations on data. Additionally, there are client layers where all data management operations take place.

To receive actionable results from analytical queries, data must be transformed into the most palatable formats. Different data processing options are available for that purpose. The computational and analytical tasks performed by a corporation, as well as the resources available, may influence the choice of the best strategy.

Processing environments are divided into centralized and distributed categories according to the use of one or more machines:

- With **centralized processing**, everything is done on a single computer system, as the name suggests (a dedicated server that hosts all data). As a result, numerous users simultaneously share the same resources and data. A single point of failure also coexists with the single point of control. The entire system fails if anything goes wrong, not just a small portion of it.

- When datasets are just too big to process on a single machine, **distributed processing** is used. With this method, big datasets can be divided into manageable chunks and stored among various servers. Consequently, parallel data processing is made possible. The beautiful thing about the distributed strategy is that if a network server crashes, data processing activities can be shifted to other available servers.

- There are batch and real-time processing methods distinguished based on the processing time:

- **Batch processing** is a technique that involves processing groups of data at a time that have been collected over a period of time. This occurs when computing resources are easily accessible, saving on them but adding time to the batch task completion process. When accuracy rather than speed is the priority, batch processing may be preferred to real-time processing.

- Due to the constant input, transformation, and output of data items, **real-time processing** makes sure that the data is constantly up to date. According to this processing style, all computational tasks are completed quickly, typically in a matter of seconds or milliseconds. Consider fleet management software that continuously monitors a vehicle's location and travel path. Real-time processing, however more difficult to implement, is an excellent choice for making decisions more quickly.

3. **Data cleansing**
Data, no matter how tiny or large, must first be completely cleaned to assure the highest quality and most accurate findings. Simply said, the data cleansing process entails checking the data for errors, duplicates,

inconsistencies, redundancies, incorrect formats, etc. This verifies the data's usability and relevance for analytics. The removal or consideration of any inaccurate or irrelevant data is required. A variety of data quality technologies can identify and fix dataset issues.

4. **Data analysis**
   This is the point at which big data becomes useful knowledge that, among other things, fuels business growth and competition. There are numerous methods and procedures for making sense of the enormous volumes of data. Here are a few of them:

   - **Natural language processing (NLP)** is a technique that enables computers to comprehend and act upon spoken or written human language.

   - **Text mining** is a cutting-edge analytical technique used to interpret big data that is presented in textual formats including emails, tweets, research papers, and blog posts.

   - The process of identifying data points and events that differ from the rest of the data is known as **outlier analysis**, also known as anomaly detection. It is frequently used in fraud detection operations.

   - **Analyzing sensor data** involves looking at the data that is constantly produced by various sensors that are put on actual objects. When completed promptly and correctly, it can assist in providing a complete picture of the equipment state as well as identifying problematic behavior and foreseeing problems (Table 4.1).

Cloud computing versus big data analytics [6]

How the cloud can help transform big data? [11]

The cost of maintaining and scaling legacy database management and data warehouse equipment is rising. Additionally, they are unable to handle the difficulties posed by today's unstructured data, Internet of Things (IoT), streaming data, and other key technologies for digital transformation.

The cloud holds the solution to massive data transformation. The majority of IT specialists working on big data decisions have already expanded their use of cloud computing or have already done so.

Three main factors – scalability (growing capacity), speed (providing findings more quickly), and security – transform large data (securely storing information). These advantages give businesses greater agility and the ability

Table 4.1 Differences between cloud computing and big data analytics [6].

Cloud computing	Big data analytics
It is used to store data on remote server across the world.	It is used to analyze voluminous data to draw conclusions important for decision-making.
It is related to computational devices.	It is an analysis of very large and complex data.
It aims to provide computational services.	It is useful for getting insights into decision-making.
Services provided by cloud computing include: IaaS, PaaS, and SaaS.	Some of the solutions for big data analytics include: Hadoop, MapReduce, Hive, etc.
It comprises a network of servers connected through the Internet to provide the services.	The data used is stored either on a cloud or in an on-premise data center.
It provides centralized access for auto-scaling, and on-demand services.	It uses distributed framework with parallel computing.
It is not only used for storing large datasets but is also useful in providing a means to analyze them.	Data is cleaned, made structural, and interpreted here.
Examples of CSPs include AWS, Microsoft, Google, etc.	Distributed data in different computing is analyzed by using Cloudera, Apache, etc.

to react swiftly to shifting market conditions while yet maintaining control over critical corporate assets.

Relationship between cloud and big data [9]

It is difficult to store, organize, and analyze this data. Big data and cloud computing work together to address this issue. The following are the ways in which this problem can be solved using these two technologies:

Storage: The storage of huge data is one of the main issues. Physical infrastructure is insufficient to appropriately store this enormous volume of data. Even if capacity is not a problem, customers may still have problems due to the physical storage's scalability. To store and retrieve huge data, cloud computing offers dependable, secure, and scalable storage facilities. Because of the decentralization and elimination of physical infrastructure, these remote storages relieve users of maintenance duties. Scalability is not a problem because cloud storage services are based on a pay-as-you-go basis and this storage may be readily raised or lowered as per users' needs.

Accessibility: All of the virtual services offered by cloud services—SaaS, IaaS, or PaaS—are hosted by external entities. Without installing and executing the program, users can access them from web browsers and make changes accordingly. The quick conveyance of data over numerous channels without an external source goes hand in hand with the simplicity of accessibility. Consider a Google Docs document as an example. It is kept on the cloud as opposed to the papers that are kept on your machine. You need only copy the URL and send it to send or transfer this file.

Security: In today's information technology industry, data security is a major concern. In 2020, there were 1001 instances of data breaches in the US alone, according to Statista. Cloud services are available and open-sourced. As a result, secure storage is difficult. Depending on the demands of the customer, cloud services offer different levels of security. Customers might ask for complex security features like data hiding, logging, encryption, etc., or they can just want their data protected through basic logical access. A service level agreement, a contract between users and service providers, is involved. This agreement has provisions for data protection, security, accessibility, capacity modification, and scalability. Cloud computing makes it simpler to store, manage, and access huge data in this way. The demand for cloud computing will only grow as electronic gadgets become more widely used.

Cloud applications for big data [8]

IaaS in the public cloud: Access to nearly infinite storage and computational capacity is provided when using a CSP's infrastructure for big data analytics.

Enterprise customers can use IaaS to build affordable and scalable systems, with CSPs taking on the challenges and costs of managing the infrastructure.

Instead of buying, installing, and integrating hardware themselves, a company customer can use the cloud resource as-needed if the size of their operations changes or they want to expand.

PaaS in the private cloud: PaaS suppliers are starting to combine technologies like MapReduce into their PaaS services, which removes the need to manage several software and hardware components.

Web developers, for instance, can employ unique PaaS environments throughout the entire development, testing, and hosting process for their websites. Platform as a service can be used by companies who are creating their own internal software, notably to establish separate development and testing environments.

SaaS in the hybrid cloud: Many businesses feel the need to listen to their clients, particularly on social media. Vendors of SaaS offer both the analysis platform and the social media data.

The best illustration of a SaaS application in business is office software. SaaS can be used to carry out tasks in the areas of planning, sales, invoicing, and accounting. Businesses may prefer to utilize a single piece of software for each of these functions, or a number of software packages that each handle a separate function.

A username and password are required to access the software online from a device connected to the office after getting the subscription. They can switch to software that more effectively meets their needs if necessary.

Examples of big data analytics in the cloud [5]

Scalability, fault tolerance, and availability are required for the storage and processing of large volumes of data. All of these are provided by cloud computing via hardware virtualization. Big data and cloud computing are thus complementary ideas since the cloud makes big data accessible, scalable, and fault tolerant. Big data is viewed by businesses as a valuable commercial prospect. As a result, numerous new businesses have begun to concentrate on providing Big Data as a Service (BDaaS) or DataBase as a service, including Cloudera, Hortonworks, Teradata, and many others (DBaaS). Customers can access big data on demand through companies like Google, IBM, Amazon, and Microsoft. We then give two examples – Nokia and RedBus – that illustrate how large data is effectively used in cloud contexts.

Nokia: One of the first businesses to recognize the value of big data in cloud systems was Nokia (Cloudera, 2012). The business previously employed different DBMSs to satisfy each application demand. The organization chose to switch to Hadoop-based systems, combining data within the same domain, and utilizing the use of analytics algorithms to acquire accurate insights into its clients after discovering the benefits of doing so. Hadoop costs less per terabyte of storage than a conventional RDBMS because it employs commodity hardware.

Incorporating the most well-known open source projects in the Apache Hadoop stack into a single, integrated package with stable and dependable releases, Cloudera Distributed Hadoop (CDH) represents a fantastic opportunity for implementing Hadoop infrastructures and shifting IT and technical concerns onto the vendor's specialized teams. Nokia saw Big Data as a Service (BDaaS) as a benefit and relied on Cloudera to quickly establish a Hadoop environment that meets its needs. Nokia

Figure 4.12 Cloudera data platform [13].

was greatly assisted by Hadoop and, in particular, CDH, in meeting their requirements (Figure 4.12).

RedBus: The biggest business in India that specializes in online hotel and bus ticket booking is called RedBus. This business sought to adopt a potent data analysis tool to learn more about its bus reservation service. Its datasets are easily expandable to two terabytes. The software would need to be able to analyze inventory and booking information from hundreds of bus companies operating more than 10.000 routes. The business also needed to avoid establishing and maintaining a complicated internal infrastructure. RedBus first contemplated installing internal Hadoop server clusters to process data.

However, they soon came to the realization that setting up such a system would take too much time, and that the infrastructure would need to be maintained by specialized IT teams. The business saw Google bigQuery as the ideal solution for their requirements at that point, enabling them to:

- Know the number of times customers attempted to find a seat but were unsuccessful due to bus overload

- Examine decreases in bookings

- Immediate identification of problems related to servers by analysis of data regarding server activity

Google BigQuery

Figure 4.13 Cloudera data platform [12].

Business benefits came from RedBus' shift to big data. RedBus now has access to real-time data analysis tools thanks to Google bigQuery, which is 20% less expensive to maintain than a sophisticated Hadoop system. Examples from Nokia and RedBus demonstrate how firms can acquire a competitive edge by embracing big data. Additionally, BDaaS offered by big data suppliers enables businesses to concentrate on their core business requirements while leaving the technological intricacies to big data providers (Figure 4.13).

Advantages of big data analytics in cloud
Agility
Early data management and storage infrastructure were much slower and more difficult to administer. Installing and running a server would take weeks of time. Cloud computing is currently available and can give your business all the resources it needs. With the help of a cloud database, your business may quickly and effortlessly set up hundreds of virtual servers.

Affordability
For a business that is on a tight budget but yet wants to have updated technology, cloud computing can come to its rescue. By choosing what they want, businesses can pay for it as they go. The resources needed to manage big data are inexpensive and easily accessible. Prior to the advent of the cloud, businesses would spend a significant amount of money establishing IT staff and further funds maintaining the gear. Companies can host their big data on remote servers or only pay for the power and storage space they use each hour.

Data processing
Processing data becomes a problem as a result of the data explosion. The unstructured, chaotic data produced by social media alone, such as tweets, posts, images, videos, and blogs, cannot be handled under a single category. Data can be analyzed from both structured and unstructured sources using

big data analytics platforms like Apache Hadoop. Small, medium-sized businesses as well as bigger ones can access and streamline the process thanks to cloud computing.

Feasibility

The virtual aspect of the cloud enables virtually limitless resources on demand, whereas conventional systems would require the addition of more physical servers to the cluster to boost computing power and storage capacity. With the cloud, enterprises can quickly scale up or down with ease to the desired level of processing power and storage space.

Large datasets now need to be processed differently due to big data analytics. The cloud environment is the ideal platform to carry out this activity, and demand for processing this data might increase or decrease at any time of the year. Since the majority of solutions can be found in SaaS models on the cloud, no additional infrastructure is required.

Challenges to big data in the cloud

Terabytes of data have been made available to enterprises thanks to big data, but handling this data within a conventional and fixed framework has proven to be a challenge. The problem is analyzing vast amounts of data to get only the most pertinent pieces. The challenge of analyzing these vast amounts of data frequently becomes challenging. Moving enormous volumes of data and giving the information needed to access it are problems in the era of high speed connectivity. Data security issues arise because these vast amounts of data frequently contain sensitive information.

Cloud security concerns are therefore a concern for both organizations and cloud providers. Cloud abuse, denial of service attacks, phishing attacks, and ransomware are some malicious attacks which are done by attackers on the cloud. A lot of organizations in the world have issues because of ransomware attacks. Hackers may obtain important information from insecure APIs or some system entry points. Governments also enforce rules that necessitate the proximity of data centers to users as opposed to providers.

Implementing cutting-edge technology that can identify problems before they worsen is the best way to overcome these obstacles. Data encryption and fraud detection are crucial in the fight against attackers. People should also understand while searching for smart solutions will give good returns, they also must maintain and protect their own data properly.

4.3 Conclusion

Big Data analytics is very much needed in recent times because we are generating voluminous amounts of data which is unstructured every second. To extract data of value from such complex datasets is required for better deci-sion-making and to understand consumer behavior for better provision of ser-vices. The cloud is very flexible and therefore it is an ideal medium through which we can perform big data analytics. With the increasing demand for cloud computing services by companies because of its versatility, it has become much more economical for organizations to store and perform analy-sis using the cloud. The cloud makes data integration easier for companies as it helps in aggregating data from numerous sources. The cloud also ensures better security, availability, and accessibility for customers. With these two technologies combined, companies can speed up their development and adapt to the current market conditions while ensuring customer satisfaction.

References

[1] https://www.cse.iitb.ac.in/~abhirup09/Docs/cloud_computing_final_report.pdf

[2] https://www.esds.co.in/blog/cloud-computing-types-cloud/

[3] https://www.geeksforgeeks.org/types-of-cloud/

[4] https://www.jigsawacademy.com/blogs/cloud-computing/characteristics-of-cloud-computing/

[5] https://acme.able.cs.cmu.edu/pubs/uploads/pdf/IoTBD_2016_10.pdf

[6] https://www.geeksforgeeks.org/difference-between-cloud-computing-and-data-analytics/

[7] https://www.firmex.com/resources/blog/7-big-data-techniques-that-create-business-value/

[8] https://data-flair.training/blogs/big-data-and-cloud-computing-comprehensive-guide/

[9] https://itchronicles.com/big-data/the-roles-and-relationships-of-big-data-and-cloud-computing/

[10] https://www.altexsoft.com/blog/big-data-analytics-explained/

[11] https://www.techtarget.com/searchcio/Rackspace/4-Steps-to-Big-Data-Transformation-in-the-Cloud

[12] https://usercentrics.com/knowledge-hub/deal-with-google-bigquery/

[13] https://docs.nvidia.com/cloudera/reference-architecture-nvidia-cloudera-components.html

[14] http://www.sun.ac.za/english/data-science-and-computationalthinking/summer-school/social

[15] https://www.expressanalytics.com/blog/social-media-sentiment-analysis/

[16] https://towardsdatascience.com/apriori-algorithm-for-association-rule-learning-how-to-find-clear-links-between-transactions-bf7ebc22cf0a

[17] https://medium.com/@priyankaparashar54/decision-tree-classifica-tion-and-its-mathematical-implementation-c27006caefbb
[18] https://corporatefinanceinstitute.com/resources/knowledge/finance/regression-analysis/
[19] https://www.analyticsvidhya.com/blog/2021/04/machine-learning-basics-for-data-science-enthusiasts/
[20] Prem Chander, K., Sharma, S. S. V. N., Nagaprasad, S., Anjaneyulu, M., & Ajantha Devi, V. (2020). Analysis of Efficient Classification Algorithms in Web Mining. In *Data Engineering and Communication Technology* (pp. 319–332). Springer, Singapore.
[21] https://towardsdatascience.com/an-introduction-to-genetic-algorithms-c07a81032547#:~:text=A%20genetic%20algorithm%20is%20a,mutation%2C%20selection%2C%20and%20crossover.
[22] https://www.educative.io/answers/what-is-amazon-ec2
[23] https://blog.lawrencemcdaniel.com/setting-up-aws-s3-for-open-edx/
[24] https://medium.com/@sspatil/introduction-to-cloud-computing-4c8e-8ce83a7f
[25] Krishnaveni, N., Nagaprasad, S., Khari, M., & Devi, V. A. (2018). Improved Data integrity and Storage security in Cloud Computing. *International Journal of Pure and Applied Mathematics*, *119*(15), 2889–2897.

5

Big Data Analytics with Artificial Intelligence: A Comprehensive Investigation

Sushma Malik[1], Anamika Rana[2], and Monisha Awasthi[3]

[1]Institute of Innovation in Technology and Management, India
[2]Maharaja Surajmal Institute, India
[3]School of Computing Sciences, Uttaranchal University, India

Email: sushmamalikiitm@gmail.com; anamica.rana@gmail.com;
monishaawasthi2011@gmail.com

Abstract

The amount of digital data is expanding rapidly because of the quick advancements in digital technologies. As a result, several sources, including social media, smartphones, sensors, etc., produce a lot of data. Emerging technologies like the Internet of Things (IoT) and recent developments in sensor networks have allowed for the collection of vast amounts of data. Such vast kinds of data that cannot be stored and processed by traditional relational databases and analytical methods are called big data. More effective techniques with high analytical accuracy are required for the investigation of such vast amounts of data. It is consequently necessary to develop fresh tools and analytical methods to find patterns from massive datasets. Big data is swiftly created from a variety of sources and formats. To effectively leverage quickly changing data, novel analytical methods must now be able to identify correlations between them. In big data analytics, artificial intelligence (AI) techniques like machine learning, knowledge-based, and decision-making algorithms can produce results that are more accurate, quicker, and scalable. Despite this interest, we are not aware of any comprehensive analysis of the various artificial intelligence algorithms for big data analytics. The main objective of the current survey is to examine artificial intelligence-based big data analytics research. Well-known databases like ScienceDirect, IEEEXplore, ACM Digital Library, and SpringerLink are used to choose relevant research articles. "Big data Analytics", "Artificial intelligence", "Big data Analytics" and "Machine Learning", keywords are used to search the related research papers for the period 2017–2022. The AI techniques which are used to investigate the research papers are knowledge based, machine learning, search methods, and decision-making categories. In each category, several articles are investigated. This chapter also compares the selected AI-driven big data analytics techniques in terms of scalability, precision, efficiency, and privacy.

5.1 Introduction

The amount of digital data is rapidly expanding due to the rapid advancement of digital technologies. As a result, several sources, including social networks, smartphones, sensors, etc., produce a lot of data. Big data is the practice of storing such enormous volumes of data that traditional relational databases and analytical approaches are unable to hold. It is consequently necessary to develop fresh tools and analytical methods to find patterns from massive datasets. Big data is swiftly created from a variety of sources and formats. To effectively leverage quickly changing data, novel analytical methods must now be able to identify correlations between them.

Before the big data revolution, businesses were unable to keep all nor effectively manage enormous amounts of material in their archives sets. The storage capabilities of older technologies are constrained. Strict management tools cost a lot of money. They are not scalable; Performance and adaptability are required in the context of big data. Big data management necessitates substantial resources, novel techniques, and advanced technology. More specifically, big data needs to be cleaned, and huge data have to be processed, analyzed, secured, and given granular access to expanding sets of data. Businesses and sectors are more conscious that data analysis is becoming a more important component of competitiveness, new insight discovery, and service personalization [1].

Traditional processing methods struggle to handle a large quantity of data. Effective methods for data analysis in large data issues must be developed. Numerous big data frameworks, such as Hadoop and Spark, have made it possible to distribute and analyze a lot of data. To produce quicker and more accurate answers for huge data analytics, various types of Artificial intelligence (AI) approaches, such as machine learning (ML) and search-based methods, were also introduced. Big data analysis now has new prospects thanks to the integration of AI methods with big data tools.

In the remaining sections of the article, the classification that follows will be covered. "Context and Related Work" has a review of the prior studies. We discussed the method of article selection in "Selection Method for research". In "Mechanisms for Big data analysis driven by AI", the planned taxonomy for the chosen big data analysis research and the selected studies are examined. In "Results" the studies that were investigated will be compared. Finally, in the sections titled "Open Issues and Challenges" and "Conclusion", respectively, some unresolved issues and a conclusion are presented.

5.2 Context and Related Work

5.2.1 Big data

We currently live in a data-driven world. The only thing we can see is data. Therefore, how to store data and how to process it are crucial. Exactly to say, what is big data? Big data is the term used to describe information that cannot be processed or stored using current technology. Traditional database technology does not allow for the storage, processing, or analysis of enormous amounts of data. Big data has an elusive character and requires a variety of techniques to transform the data into fresh insights [2].

Big data refers to the enormous computing power required to handle the complexity and volume of data coming from numerous sources, including the Internet and remote sensor networks.

The term "big data" refers to a collection of methods and tools that use novel integration approaches to glean valuable insights from vast, intricate, and scale-increasing data sets [2].

Huge amounts of data in a variety of heterogeneous formats were gathered from numerous sources including sensors, transactional apps, and social media. Big data is a term used to describe a rising body of data that includes structured, unstructured, and semi-structured data in a variety of formats. Such a large amount of diverse data cannot be processed by current database management systems (DBMSs). Therefore, to process huge data, strong technologies, and cutting-edge algorithms are required [3].

Big data can be characterized by various (Figure 5.1) Vs, including volume, velocity, variety, and veracity [4].

- **Volume**: This alludes to the enormous amounts of data generated every second. Big data frameworks can process these enormous amounts of data. Millions of applications and gadgets (ICTs, smartphones, product codes, social media, sensors, logs, etc.) constantly produce large amounts of digital data.

Figure 5.1   Characteristics of big data.

- **Velocity**: This indicates how quickly data is produced and processed to get insightful information. Data should be processed quickly to obtain important information and pertinent insights because they are created quickly.

- **Variety**: This details the different data formats, including documents, movies, and logs.

- **Veracity**: This shows the variables affecting data quality. In other words, it describes any biases, noise, abnormalities, etc. in the data.

Today, extra "Vs" and other traits like "value" and "volatility" have been added to the definition of big data.

Massive data management is crucial for effectively managing big data and producing high-quality data analytics. It involves effective data gathering from multiple sources, effective data storage utilizing a range of techniques and technologies, effective data cleansing to remove errors and format the data consistently, and effective data encoding for security and privacy. This procedure aims to make sure that trustworthy data is managed effectively, stored efficiently, and kept safe [1].

5.2.2 Applications of big data

Large quantities of complicated, raw data are referred to as big data. By enabling data scientists, analytical modelers, and other professionals to analyze enormous volumes of transactional data, today's businesses employ big data to make businesses more informed and enable business decisions. The precious and potent fuel that powers the enormous IT firms of the twenty-first century is known as big data. Big data is a technology that is gaining traction across many industries. In this section, we will talk about big data applications [2–4].

5.2.2.1 Internet of Things (IoT)

One of the key markets for big data applications is the Internet of Things. The Internet of Things (IoT) applications are constantly changing due to the wide diversity of things. There are several big data applications available now that serve logistical businesses. Sensors, wireless adapters, and GPS can be used to track the locations of moving objects. As a result, these data-driven solutions give businesses the ability to improve delivery routes in addition to managing and supervising their workforce. By utilizing and synthesizing different information, including prior driving experience, this is accomplished. Based on the use of IoT data, smart cities are another popular study topic.

5.2.2.2 Transportation and logistics

RFID (radiofrequency identification) and GPS are widely used by public transportation companies to track buses and investigate valuable data to enhance their offerings.

5.2.2.3 Smart grid

It is essential to manage and track the national electronic power usage in real-time. Multiple connections between smart meters, sensors, control centers, and other infrastructures are used to achieve this. Big data analytics aids in the detection of anomalous activities in linked devices and the identification of transformers that are at risk. Thus, Grid Utilities can decide on the most effective course of action. Modeling event situations is possible because of the real-time analysis of the generated big data. This makes it possible to create strategic preventative plans to lower the expenses of corrective action (Figure 5.2).

5.2.2.4 E-Health

Personalizing health services already happens with connected health platforms. Various heterogeneous sources (such as laboratory and clinical data, patient symptoms uploaded through remote sensors, hospital activities, and pharmaceutical data) are used to create big data. The enhanced analysis of medical data sets has many beneficial applications. It makes it possible to customize health services (for instance, clinicians can track online patient symptoms to change prescriptions); to modify public health policies by population symptoms, the progression of diseases, and other factors. It can also be used to reduce spending on healthcare and improve hospital operations [5].

Figure 5.2   Applications of big data.

5.2.2.5 Tracking customer spending habit and shopping behavior

Management teams of large retailers (such as Amazon, Walmart, Big Bazar, etc.) are required to maintain data on customer spending habits (including the products they purchase, the brands they prefer, and how frequently they do so), shopping patterns, and their favorite products (so that they can keep those products in the store). Based on data about which product is most frequently searched for or purchased, the production or collection rate of that product is fixed.

5.2.2.6 Recommendation

Big retail stores give recommendations to customers by studying their purchasing patterns and shopping behavior. Product recommendations are made on e-commerce sites like Amazon, Walmart, and Flipkart. They keep track of the products that customers are looking for and, using that information, recommend that kind of goods to them.

5.2.2.7 Secure air traffic system

Sensors are present at various points of flight (such as propellers and other objects). These sensors record information on environmental conditions, including temperature, moisture, and flying speed. An environmental parameter for flight is set up and adjusted based on such data analysis. The length of time the machine can function flawlessly before needing to be replaced or repaired can be calculated by examining the machine-generated data from the flight.

5.2.2.8 Virtual personal assistant tool

Big data analysis enables virtual personal assistant tools (like Siri on Apple devices, Cortana on Windows, and Google Assistant on Android) to respond to a variety of customer questions. This program keeps note of the user's location, local time, season, and other information about the query asked, etc. It offers a solution after analyzing all this data.

5.2.2.9 Education sector

Organizations running online educational courses use big data to find candidates who are interested in taking the course. If someone searches for a YouTube tutorial video on a certain topic, an organization offering online or offline courses on that topic will subsequently send that individual an online advertisement for their course.

5.2.2.10 Sector of media and entertainment

Companies that offer media and entertainment services like Netflix, Amazon Prime, and Spotify analyze the data they receive from their customers. To determine the next business strategy, information is gathered and evaluated about the types of videos and music consumers are watching and listening to, how long they spend on a website, etc.

5.2.3 Big data challenges

Big data mining presents a wide range of alluring possibilities. However, there are several difficulties that researchers and professionals must overcome when examining big data sets and gleaning knowledge and value from these informational gold mines. Data collection, storage, searching, sharing,

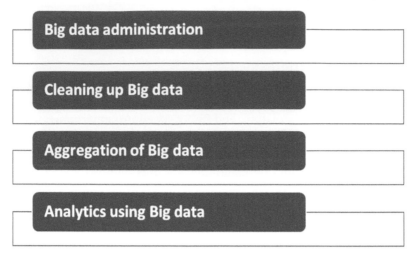

Figure 5.3 Challenges of big data.

analysis, management, and visualization are just a few of the areas where challenges exist. In addition, security and privacy concerns exist, particularly in remote data-driven applications.

This section goes into greater detail about various technical problems that need further investigation (Figure 5.3).

5.2.3.1 Big data administration

The obstacles that data scientists face while working with big data are numerous. How to gather, combine, and store massive data sets produced by diverse sources while requiring less technology and software is one difficulty.

5.2.3.2 Cleaning up big data

These five phases (cleaning, aggregation, encoding, storage, and access) are common in traditional data management and are not brand-new. The difficulty in processing big data in a distributed setting with a range of applications is managing its complexity (velocity, volume, and variety). It is crucial to confirm the validity of sources and the caliber of the data before utilizing resources to produce reliable analytical results. Data sources, however, might have inaccuracies, noise, or insufficient data. How to clean up such massive data sets and determine which data is trustworthy and helpful presents challenges.

5.2.3.3 Aggregation of big data

Synchronizing external data sources and distributed big data platforms (such as applications, repositories, sensors, networks, etc.) with an organization's internal infrastructures is another difficulty. Analysis of the data produced within businesses is frequently insufficient. Going a step further and combining internal data with external data sources is crucial to gain insightful knowledge.

5.2.3.4 Analytics using big data

Big data presents unprecedented difficulties in utilizing such vast, growing volumes of data, but it also offers enormous benefits and the potential to alter many industries. To comprehend the links between features and examine data, advanced data analysis is required.

5.2.4 Big data platforms

In this section, we included the platforms and tools of Big Data which are used for the huge amount of data. A big data platform is an integrated IT solution for managing big data that integrates multiple programs, software tools, and hardware to provide easy-to-use tool systems for businesses. Big data platforms are used for the development, implementation, and management of big data. There are many open-source and for-profit big data platforms on the market with a wide range of features that can be used in big data environments.

5.2.4.1 Big Data Platform Features

The following are the key characteristics of a good big data analytics platform [1]:

- According to the needs of the business, the big data platform should be able to handle new platforms and tools. Because company requirements may alter because of new technology or changes in business procedures.

- It ought to accommodate linear scale-out.

- It ought to be able to deploy quickly.

- It ought to support many data formats.

- A platform should offer capabilities for reporting and data analysis.

- It ought to offer real-time data analysis tools.

- Tools for searching across enormous data sets should be available (Table 5.1).

Table 5.1 Big data platform

Big data platform	Description
Hadoop	Hadoop is an open-source, Java-based server and programming framework that is used in a clustered environment to store and analyze data using hundreds or even thousands of commodity machines. Large datasets may be stored and processed quickly and fault-tolerantly with Hadoop. HDFS (Hadoop File System) is used by Hadoop to store data on a cluster of inexpensive machines.
Cloudera	One of the earliest for-profit Hadoop-based big data analytics platforms to offer big data solutions is called Cloudera. These technologies all offer real-time processing and analyses of enormous data sets and are built on the Apache Hadoop framework.
Amazon Web Services	As a component of their Amazon Web Services offering, Amazon provides a Hadoop environment on the cloud. Elastic cloud compute and simple storage service from Amazon power the hosted AWS Hadoop solution (S3).
Hortonworks	Without any proprietary software, Hortonworks only uses open-source software. Support for Apache HCatalog was first incorporated by Hortonworks. Big data business Hortonworks is situated in California. This business creates and supports Apache Hadoop applications.
MapR	Another big data platform, MapR, handles data using the Unix filing system. This technique is simple to learn for anyone who is familiar with the Unix operating system and does not need HDFS. This solution combines real-time data processing with Hadoop, Spark, and Apache Drill.
Open platform by IBM	In addition, IBM provides the Hadoop eco-system-based big data platform. IBM is a reputable name in software and data processing. It takes advantage of the newest Hadoop program.
Windows HDInsight	A commercial big data platform from Microsoft, HDInsight is based on the Hadoop distribution as well. This is the main Hadoop distribution product, and it works with Windows and Azure.

5.3 Artificial intelligence

The study of artificial intelligence focuses on developing computer programs that can carry out intelligent tasks that previously required humans alone. AI is the general term for Artificial intelligence. It trains computers to learn human skills like learning, judgment, and decision-making and employs computers to emulate intelligent human behavior. The goal of AI is to simulate human intellectual activities by taking knowledge as the object, acquiring knowledge, studying and analyzing how knowledge is expressed, and applying these methods.

To succeed in the new international competition, many countries have introduced preferential treatment policies and strengthened the deployment of critical technologies and skills. Big companies like Google, Microsoft, and IBM are devoting themselves to artificial intelligence and using it in more and more fields. Artificial intelligence has become a research center for science and technology. To integrate cognition, machine learning, emotion recognition, human-computer interaction, data storage, and decision-making, AI is a multidisciplinary technology. John McCarthy made the initial suggestion at the Dartmouth Conference in the middle of the 20th century. The events in the AI chronology are covered in the following table (Tables 5.2–5.4) [6–9].

5.3.1 Examples of AI applications

Based on the relatively mature growth of technological factors like data, algorithms, and computer power, AI has started to effectively generate economic benefits and solve real-world problems. Finance, healthcare, the automobile, and retail sectors, for example, have relatively developed AI application scenarios from an application standpoint.

5.3.1.1 The automotive industry

Autonomous driving is the result of the comprehensive integration of modern information technologies, including automotive, artificial intelligence (Figure 5.4), and the Internet of Things. For current international travel, as well as for contacts and travel information, this is a crucial direction.

Autonomous driving uses sensors like lidar and other sensors to gather information about pedestrians and road conditions in order to continuously improve and, ultimately, deliver the best route and control plan for moving cars on the road [8].

5.3.1.2 The financial markets

The market for AI in the finance industry is expanding. It can forecast risks and the direction of the stock market using technical methods like machine learning. Financial institutions use machine learning techniques to integrate different data sources, manage financial risks, and give customers real-time risk warning statistics. Big data is used by financial institutions to analyze

Table 5.2 Advancements in AI [13].

Period	Events and Details
Founding period	At a Dartmouth University academic symposium in 1956, several eminent scientists debated the creation of artificial intelligence. McCarthy was the one who initially coined the phrase "artificial intelligence," giving rise to the concept.
The first golden age	The number of universities that created AI laboratories in the early stages of the field and obtained R&D support from governmental organizations. Feigenbaum first put forth the idea of knowledge engineering in the late 1970s. Expert systems have quickly advanced, and their applications have yielded enormous advantages. But as several issues began to surface, including the challenge of learning from expert systems, artificial intelligence hit its low point.
The second golden period	The proposals of the Hopfield neural network (1982) and the BT training algorithm led to a flourishing in the development of AI, including speech recognition and translation. In the late 1990s, however, many believed that social interaction and AI were still decades away. As a result, artificial intelligence experienced another downturn around the year 2000.
The third golden period	The period from 2006 to the present has seen a rapid advancement in artificial intelligence. The widespread use of GPUs, which allows for quicker and more potent parallel computing, is mostly to blame for the quick progress. Another factor is the storage space's limitless extension, which enables widespread access to material like maps, photos, text, and data information.

associated financial risks, offer pertinent financial assets with real-time risk warnings, conserve human, and material resources for investment and financial management, establish a logical and scientific risk management system, and lay the foundation for the expansion of the business.

The health industry
AI-related algorithms are utilized in the medical industry to create new medications, diagnose cancer, and provide medical aid. AI algorithms were used in the analysis and processing to provide medical assistance to stakeholders and make medical diagnoses more efficiently and accurately [10].

Table 5.3 Artificial intelligence enablers and technologies.

Driver or Technology	Explanation
Algorithms	Any operation must be carried out in a specific order. Algorithms are the techniques and procedures used to address the issue.
Big data	Big data is a term used to describe a collection of data that cannot be acquired, managed, and processed in a certain amount of time using traditional computing methods. Big data necessitates the development of new processing models with improved insight and discovery, process optimization, and varied information asset capabilities.
Machine Learning	The study of artificial intelligence is known as machine learning. Artificial intelligence is the primary study topic in this area, particularly how to enhance the effectiveness of algorithms for empirical learning. The study of computer algorithms that can automatically get better with practice is known as machine learning. Machine learning enhances the performance requirements of computer programs using data or experience.
National Language Processing (NLP)	Natural language processing (NLP) research combines linguistics, computer science, and artificial intelligence. It focuses on the interaction between computers and human (natural) language. Machine translation represents the oldest study in natural language processing.
Hardware	A computer system's multiple physical devices made up of electrical, mechanical, and optoelectronic components are collectively referred to as hardware. These tangible objects serve as a material foundation for the operation of computer software and, following the system structure's needs, constitute annuities. GPUs make up most of the deep learning hardware platform.
Computing Vision	How to make machines "see" is the subject of the study known as computer vision. Additionally, it describes the process of identifying, tracking, and measuring objects without the use of human sight by using cameras and computers. Through this processing, a picture is made more suited for transmission to an inspection tool or for human observation.

Table 5.4   The areas of AI research.

Fields	Explanation
Expert System	A computer intelligence program system called an expert system has specialized knowledge and experience. It uses AI data representation and data inference techniques to simulate complex problems that experts often deal with, simulating the problem-solving skills of human experts. It can have problem-solving abilities on par with those of an expert.
Machine Learning	Probability theory, statistics, approximation theory, convex analysis, algorithm complexity theory, and other fields are all involved in machine learning. The study of how computers mimic or actualize human learning patterns to pick up new information or abilities, as well as how they restructure the current knowledge structure to continuously enhance their performance, is known as machine learning.
Robotics	The science of robotics is concerned with the creation, use, and application of robots. It primarily examines the connection between robot control and object processing.
Decision Support System	The term "knowledge-intelligence" and the management science subfield in which decision support systems fall are closely related.
Pattern Recognition	The subject of pattern recognition focuses on how to provide machines with perceptive abilities. It focuses mostly on the identification of visual and aural patterns.

5.3.1.3 The retailing industry

True unmanned retail is achieved in the retail sector by offline physical retail establishments using AI, which lowers costs and significantly boosts productivity. By incorporating AI into the recommendation system, it will be possible to improve market forecasting, boost online sales, and save inventory costs. Many e-commerce sites have implemented the recommendation system, which generates an e-commerce suggestion model based on a user's expected preferences [9].

5.3.1.4 The media industry

Media companies can create user-requested content with only one click using content transmission robots and brand communication robots, which can publish up to 10,000 articles per minute. This artificially intelligent media platform can analyze media delivery and delivery rules, investigate current hot events, popular opinion, and public relations marketing content all work together to create content that readers want to read.

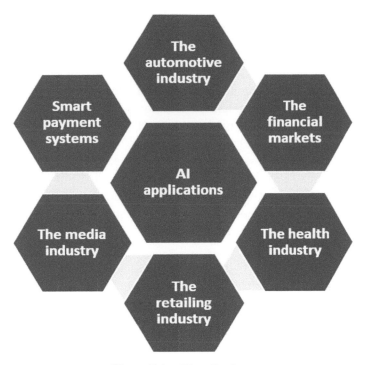

Figure 5.4 AI applications.

Smart payment systems
Many people now purchase without cash since it has become a habit. The payment can be conveniently made using a mobile device. Voiceprint payment and face scanning are two more new payment options for clients who have not utilized the scan code to purchase products. When going shopping, customers do not even need to bring their wallets, scan a QR code, or enter a password.

5.3.1.5 Smart home
To establish effective residential facilities and day-to-day family affairs management systems and improve living conditions, smart houses integrate facilities relevant to daily life. Smart home includes a variety of home appliances and others that are all-inclusive and serve consumers via intelligent intercommunication. A complete smart home system consists of several distinct home items with various functionalities in addition to a single device. In a family, there are several users instead than just one. The goal of smart home systems is to efficiently and intelligently integrate people and household goods into one, self-learning system.

Education

Intelligent robots, intelligent teaching platforms, and intelligent assessment systems, which spare teachers from laborious teaching and provide human-computer collaborative teaching, are just a few of the innovative applications born from the marriage of AI and education. The intelligent network learning space is the goal of educational AI with the combination of AI and education. Through collaboration with the government, educational institutions, businesses, and other multi-party intelligent network learning channels, education will continuously enhance teaching methods and foster the development of inventive, communicative, and learning skills. The educational resources are becoming more dynamic and open, the teaching model is becoming more intelligent, and the learning style is changing to be more individualized [11, 12].

5.3.1.6 Government

The early applications of AI were in fields with abundant data resources and well-defined scenarios.

The discipline of intelligent governance is still in its infancy, but because of AI's growing popularity, it offers a wide range of potential applications in areas including virtual government assistants, intelligent conferences, robot process automation, document processing, and decision-making. Government service and efficiency will increase thanks to AI, which will also help with the labor shortage [11].

Big data analysis solutions powered by AI

Interdependence exists between big data and artificial intelligence. Even though each discipline is unique, each must exist for the other to function to the fullest. AI does use data, but the amount of information fed into the system limits how much it can study and learn from the data. Big data is the fuel that powers the most advanced artificial intelligence systems since it offers a huge sample of this information (Figures 5.5 and 5.6).

To maximize the value of all that data, however, organizations need more than just storage and management of big data. Forward-thinking businesses are using increasingly intelligent or sophisticated forms of big data analytics to wring even more value from that information as they master big data management. They are using machine learning, which can recognize patterns and give cognitive capabilities across massive volumes of data, enabling these firms to apply the next level of analytics required to extract value from their data [14].

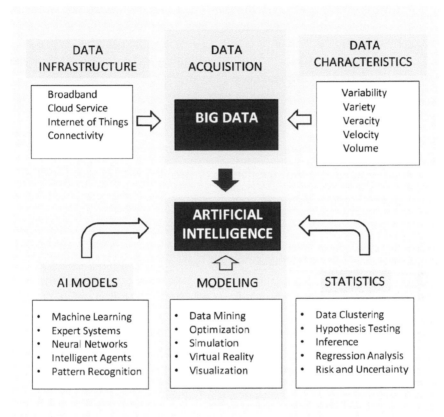

Figure 5.5 Implementation of big data and AI [13].

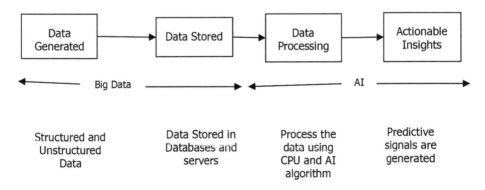

Figure 5.6 Working of big data and artificial intelligence[17].

Artificial intelligence systems may make smarter decisions, offer better user recommendations, and uncover ever-improving efficiencies in your models by utilizing large data resources. To ensure the best data is produced, a set of agreed-upon guidelines for data collecting and data organization must be in place before AI is implemented [15, 16].

A few advantages of large data with AI:

- Less time-consuming data analytics.

- Using machine learning, common data issues can be solved.

- Does not diminish the significance of people in the analysis process.

- Enhanced prescriptive and predictive analytics.

We must relate its operating concept to comprehend the relationship between AI and big data. Every search query we enter via Google Search, for instance, contains data. Like this, billions of individuals are conducting many searches on a variety of topics. These all together will make up big data. Until they are transformed into insightful knowledge, these enormous data volumes remain raw and useless.

These data sets will be examined and organized into useful insights using the proper methodology. To train machines to use the sorted data for intelligent decision-making, it should also be modeled. The computer gains greater knowledge and the capacity to carry out tasks deftly through training. The Bayes theorem states that AI systems calculate the likelihood of a future event. Using historical data analysis and machine learning to create probabilities, it is possible to anticipate future events [17, 18].

In summary, data computers are incapable of thinking for themselves. A computer can only produce intelligent results after carefully examining and interpreting data patterns.

The successful integration of AI and big data analytics has only served to advance technology. The operation of AI and Big data Analytics can be explained on several different bases [19].

The following list of variables aids in comprehending how big data analytics and artificial intelligence work.

Developing patterns

Big data analytics and AI work together to create patterns that are visible over time. In contrast to big data analytics, which focuses on data collecting, AI prefers to examine data through analysis, which ultimately aids in identifying patterns or homogeneous groupings. One of the most important roles that aid in our comprehension of how this pair operates is the ability to build patterns.

As data from big data analytics is fed to AI, the program tends to identify some patterns that have developed over time. This makes data sifting even easier

Taking anomalies out

Negating anomalies is another task that the pair performs. Any odd occurrence among the typical data is referred to as an anomaly. AI is active in recognizing unexpected information since it seeks to construct patterns, which results in the negation of anomalies. Additionally, this results in the purging of data that could otherwise be cluttered with extraneous information. While big data analytics is useful for storing data, AI takes the lead by identifying anomalies or other differences from the otherwise typical pieces of data. One of the most crucial tasks that AI and Big data Analytics jointly do when it comes to this pair's operation is negating abnormalities.

Creation of algorithms

The creation of algorithms is the confluence of AI and big data analytics' third function. Big data analytics gathers and saves data, while AI feeds off this data to develop sophisticated AI algorithms that consider the data. By identifying patterns and eliminating anomalies, AI quickly discovers and creates specific algorithms that assist it in separating the data going forward. Big data analytics has only been able to become widely used in international operations because of the technologically advanced instrument of AI.

How digital transformation is being affected by big data analytics and Artificial Intelligence

Digital transformation is a process that involves the application of technology capable of producing, storing, or processing any type of data.

The current world's digital transformation is now solely being driven by data. It has emerged as the most important resource, and without it, it would be difficult to compete in the crowded market of today. Organizations should effectively use their data because it could be a component that sets them apart in business development [20, 21].

In the modern day, data is one of the key contributors to every organization's success. Big data is being extensively used by organizations. Every day, every second, they produce and collect enormous amounts of data. It enables them to use this data to gain insightful knowledge. A company cannot just analyze a huge amount of data by hand. They use artificial intelligence to enhance their analyses and help them come to better conclusions. It can assist businesses in deriving valuable insights from the available data [13, 22, 23].

Let's look at how AI and big data are assisting businesses with their digital transformation.

Customer interaction

Big data enables to examination and produce insightful customer insights. Big data and artificial intelligence can be used to forecast client behavior. For instance, organizations have all the information on what a specific customer group does during a season that belongs to a certain region, and AI is used to display things and suggests the activities the customers enjoy.

Furthermore, organizations can use artificial intelligence to offer other solutions and resolve many of their problems with the use of this knowledge. This may eventually lead to customer retention and an improvement in their relationship with the business [24].

Improved pricing optimization and forecasting

Companies typically use information from the previous year to forecast their sales for the current year. However, forecasting and price optimization can be very challenging with traditional approaches because of several factors like shifting trends, global pandemics, or other difficult-to-predict factors. Organizations may now use big data to identify patterns and trends early on and predict how they will affect performance going forward. Supplying businesses with additional details about prospective future events with a higher possibility aids businesses in making better decisions. Businesses, particularly those in retail, can enhance seasonal forecasting by cutting errors by as much as 50% when adopting big data and AI-based methodologies.

Online/digital marketing

Digital marketing, one of the essential facets of digital transformation, is heavily relied upon by businesses. After evaluating client behavior and preferences, it aids a firm in marketing the goods and services based on the data and insights, assuring optimum engagement.

Organizations can now target audiences based on the sales and revenue for various products. Big data integration in digital marketing is assisting businesses in expanding their client base. By automating the digital marketing process, artificial intelligence is used to guide the insights from big data. AI is assisting marketers in making decisions about appropriate platforms, scheduling, content generation, and campaign reevaluation for better and better results [25].

5G

Massive connectivity across many devices is a fantastic ability of 5G. It enables the transition of big data-in-motion and big data-at-rest into some significant real-time insights with the aid of actionable intelligence and is supported by distributed architectures.

It is essential to have established communication networks that can transmit massive data and convert it into insightful information for a successful digital transformation. Here, AI is employed to ensure a seamless workflow. Artificial intelligence can choose how much data needs to be stored in case of a problem. Furthermore, when a certain network is unavailable, AI is utilized to completely change the networks.

Finding the shortcomings
A company will need to use the right data with the aid of AI if it wants to undergo a digital transformation. Integrating big data into systems is essential to obtaining the greatest benefits and identifying the operational gaps.

Data analytics, for instance, can be used to pinpoint operational problems in the world's supply and demand chains. Additionally, it can offer data and varied viewpoints on various products.

This can assist a business in overcoming early operational mistakes and improving its products considering customer demands and information gathered.

Improving cyber security
An organization's expansion and digital change make it more vulnerable to viruses and cyber-attacks. However, big data analytics can be used to pinpoint the regions that require extra security. As a result, improved security measures can be implemented following data analysis. Security systems are being automated with AI's help so they can decide which area requires extra security at any given time.

Drug discovery
By offering assessments of therapeutic compounds at the early stages of research, AI is a viable way to significantly cut the cost and time of drug discovery. Clinical and pharmaceutical data are now expanding quickly, and there is a great demand for new AI methods to handle large data sets. Recent research on deep learning modeling has demonstrated benefits over conventional machine learning techniques for this problem [26–28].

Open issues and challenges
Big Data Analytics with Artificial Intelligence (AI) has made significant advancements in recent years, but it still faces several open issues and challenges. These challenges impact various aspects of the field, from data collection and processing to model development and deployment. Here are some of the key open issues and challenges in this domain.

Building trust
AI is a new technology that is developing quickly every day. People who are not familiar with this technology often claim that it is too complicated to use. Artificial intelligence must deal with several problems related to human

trust. Human nature prefers to avoid complicated things; therefore, AI must be trustworthy and simple to use.

AI human interface
The difficulty is that many people lack data science knowledge, which prevents most people from benefiting from artificial intelligence results. Due to a lack of individuals with sophisticated capabilities in the industrial sector, business owners must teach their staff for them to take advantage of artificial intelligence's benefits.

Finance
Due to the significant costs associated with implementing artificial intelligence, many firms and organizations are unwilling to make investments in it.

Software malfunction
As we are all aware, neither humans nor technology are perfect. It can be challenging to diagnose problems with technologies when their hardware or software malfunctions (Figure 5.7).

All tasks cannot be replaced by AI
All tasks become more productive thanks to AI. This can be somewhat accurate, but not all tasks can be managed by AI. AI can manage its way into our daily lives, managing every task, every process, and even every minute. On the other hand, AI can automate all tasks and increase productivity. This feature improves the application's functionality and effectiveness.

Higher expectations
People have high expectations for AI, which is a big problem. Many people lack an understanding of artificial intelligence (AI) and, as a result, have unrealistic expectations. Humans have the propensity to have high expectations for anything popular and likely to succeed soon, yet they often fail to recognize the limitations that any technology possesses. Although AI is still in its early stages, people still have great expectations.

Security
To ensure that the data is secure and in good hands, AI and big data must acquire sensitive and crucial data from their users or clients. However, we have no idea when hackers would steal our private and sensitive information; for this reason, security is the biggest problem with any technology.

Compatibility and complexity
Big data consist of data from a number of devices with a wide range of technologies, which could result in several challenges. The proliferation of various technologies results in increasingly complicated ecosystems.

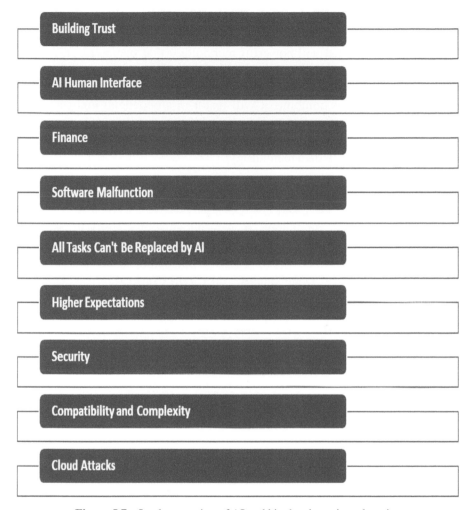

Figure 5.7 Implementation of AI and big data in various domain.

Cloud attacks

It is not surprising that dangerous infections have gained unwanted attention due to the rapid development of cloud computing technology. Big data requires a lot of data, which is kept in the cloud, increasing the danger to data security.

5.4 Conclusion

As we have shown in the studies above, integrating AI with big data has both pros and cons. This developing trend will make the Internet more useful because every implementation uses AI. Big data and AI will change many people's futures; all they require is human patience and support. Big data is a much more robust and powerful technology that perceives objects utilizing the Internet, whereas AI is relatively much wiser, more fashionable, intelligent, and quick-witted than other technologies. Therefore, if these two amazing technologies combine, it will result in a beneficial innovation and experimental technology that will benefit businesses, industries, and users by producing products that are effective and useful. All modern devices are powered by AI and big data technology.

References

[1] A. Oussous, F. Z. Benjelloun, A. Ait Lahcen, and S. Belfkih, "Big Data technologies: A survey," J. King Saud Univ. - Comput. Inf. Sci., vol. 30, no. 4, pp. 431–448, 2018, doi: 10.1016/j.jksuci.2017.06.001.

[2] T. S. Harsha, "Big Data Analytics in Cloud Computing Environment," vol. 8, no. 8, pp. 159–179, 2017.

[3] A. K. Tyagi and R. G, "Machine Learning with Big Data," SSRN Electron. J., pp. 1011–1020, 2019, doi: 10.2139/ssrn.3356269.

[4] I. E. Agbehadji, B. O. Awuzie, A. B. Ngowi, and R. C. Millham, "Review of big data analytics, artificial intelligence and nature-inspired computing models towards accurate detection of COVID-19 pandemic cases and contact tracing," Int. J. Environ. Res. Public Health, vol. 17, no. 15, pp. 1–16, 2020, doi: 10.3390/ijerph17155330.

[5] D. Singh and C. K. Reddy, "A survey on platforms for big data analytics," J. Big Data, vol. 2, no. 1, pp. 1–20, 2015, doi: 10.1186/s40537-014-0008-6.

[6] M. Mohammadpoor and F. Torabi, "Big Data analytics in oil and gas industry: An emerging trend," Petroleum, vol. 6, no. 4, pp. 321–328, 2020, doi: 10.1016/j.petlm.2018.11.001.

[7] A. M. Rahmani et al., "Artificial intelligence approaches and mechanisms for big data analytics: a systematic study," PeerJ Comput. Sci., vol. 7, pp. 1–28, 2021, doi: 10.7717/peerj-cs.488.

[8] N. Gupta and B. Gupta, "Big data interoperability in e-health systems," Proc. 9th Int. Conf. Cloud Comput. Data Sci. Eng. Conflu. 2019, pp. 217–222, 2019, doi: 10.1109/CONFLUENCE.2019.8776621.

[9] I. Antonopoulos et al., "Artificial intelligence and machine learning approaches to energy demand-side response: A systematic review," Renew. Sustain. Energy Rev., vol. 130, no. February, p. 109899, 2020, doi: 10.1016/j.rser.2020.109899.

[10] Y. Roh, G. Heo, and S. E. Whang, "A Survey on Data Collection for Machine Learning: A Big Data-AI Integration Perspective," IEEE Trans. Knowl. Data Eng., vol. 33, no. 4, pp. 1328–1347, 2021, doi: 10.1109/ TKDE.2019.2946162.

[11] R. X. Ding et al., "Large-Scale decision-making: Characterization, taxonomy, challenges and future directions from an Artificial Intelligence and applications perspective," Inf. Fusion, vol. 59, no. January, pp. 84–102, 2020, doi: 10.1016/j.inffus.2020.01.006.

[12] S. D. Erokhin, "A Review of Scientific Research on Artificial Intelligence," 2019 Syst. Signals Gener. Process. F. Board Commun. SOSG 2019, pp. 1–4, 2019, doi: 10.1109/SOSG.2019.8706723.

[13] C. Zhang and Y. Lu, "Study on artificial intelligence: The state of the art and future prospects," J. Ind. Inf. Integr., vol. 23, no. May, p. 100224, 2021, doi: 10.1016/j.jii.2021.100224.

[14] A. Esteva et al., "A guide to deep learning in healthcare," Nat. Med., vol. 25, no. 1, pp. 24–29, 2019, doi: 10.1038/s41591-018-0316-z.

[15] Y. Lu, "Artificial intelligence: a survey on evolution, models, applications and future trends," J. Manag. Anal., vol. 6, no. 1, pp. 1–29, 2019, doi: 10.1080/23270012.2019.1570365.

[16] L. Chen, P. Chen, and Z. Lin, "Artificial Intelligence in Education: A Review," IEEE Access, vol. 8, pp. 75264–75278, 2020, doi: 10.1109/ ACCESS.2020.2988510.

[17] G. Shiao, "Artificial Intelligence and Big Data." https://www.pinterest. com/pin/557179785141222187/ (accessed Sep. 29, 2022).

[18] T. H. Davenport, "From analytics to artificial intelligence," J. Bus. Anal., vol. 1, no. 2, pp. 73–80, 2018, doi: 10.1080/2573234X.2018.1543535.

[19] DataRobot, "Big data and artificial intelligence: a quick comparison - DataRobot AI Cloud," 2020. https://www.datarobot.com/blog/big-data-and-artificial-intelligence-a-quick-comparison/ (accessed Sep. 28, 2022).

[20] K. Walch, "How do Big Data and AI Work Together?," 2021. https:// www.techtarget.com/searchenterpriseai/tip/How-do-big-data-and-AI-work-together (accessed Sep. 28, 2022).

[21] S. Lincoln, "The Integration Of AI and Big Data For Digital Transformation," 2022. https://www.modernanalyst.com/Resources/ Articles/tabid/115/ID/6088/The-Integration-Of-AI-and-Big-Data-For-Digital-Transformation.aspx (accessed Sep. 28, 2022).

[22] S. Rawat, "How Big Data Analytics is Using AI? - Analytics Steps," 2021. https://www.analyticssteps.com/blogs/how-big-data-analytics-using-ai (accessed Sep. 28, 2022).

[23] Z. S. Y. Wong, J. Zhou, and Q. Zhang, "Artificial Intelligence for infectious disease Big Data Analytics," Infect. Dis. Heal., vol. 24, no. 1, pp. 44–48, 2019, doi: 10.1016/j.idh.2018.10.002.

[24] M. Obschonka and D. B. Audretsch, "Artificial intelligence and big data in entrepreneurship: a new era has begun," Small Bus. Econ., vol. 55, no. 3, pp. 529–539, 2020, doi: 10.1007/s11187-019-00202-4.

[25] Z. Allam and Z. A. Dhunny, "On big data, artificial intelligence and smart cities," Cities, vol. 89, no. November 2018, pp. 80–91, 2019, doi: 10.1016/j.cities.2019.01.032.

[26] [26]   G. Katare, G. Padihar, and Z. Qureshi, "Challenges in the Integration of Artificial Intelligence and Internet of Things," Int. J. Syst. Softw. Eng., vol. 6, no. December 2018, pp. 10–15, 2018.

[27] Devi, V. A., & Naved, M. (2021). Dive in Deep Learning: Computer Vision, Natural Language Processing, and Signal Processing. In *Machine Learning in Signal Processing* (pp. 97-126). Chapman and Hall/CRC.

[28] Salas, M., Petracek, J., Yalamanchili, P. *et al.* The Use of Artificial Intelligence in Pharmacovigilance: A Systematic Review of the Literature. *Pharm Med* 36, 295–306 (2022). https://doi.org/10.1007/s40290-022-00441-z

[29] K. Benke and G. Benke, "Artificial intelligence and big data in public health," Int. J. Environ. Res. Public Health, vol. 15, no. 12, 2018, doi: 10.3390/ijerph15122796.

[30] Q. V. Pham, D. C. Nguyen, T. Huynh-The, W. J. Hwang, and P. N. Pathirana, "Artificial Intelligence (AI) and Big Data for Coronavirus (COVID-19) Pandemic: A Survey on the State-of-the-Arts," IEEE Access, vol. 8, no. Cdc, pp. 130820–130839, 2020, doi: 10.1109/ACCESS.2020.3009328.

[31] Q. André et al., "Consumer Choice and Autonomy in the Age of Artificial Intelligence and Big Data," Cust. Needs Solut., vol. 5, no. 1–2, pp. 28–37, 2018, doi: 10.1007/s40547-017-0085-8.

[32] Y. Duan, J. S. Edwards, and Y. K. Dwivedi, "Artificial intelligence for decision making in the era of Big Data – evolution, challenges and research agenda," Int. J. Inf. Manage., vol. 48, no. January, pp. 63–71, 2019, doi: 10.1016/j.ijinfomgt.2019.01.021.

[33] H. Zhu, "Big Data and Artificial Intelligence Modeling for Drug Discovery," Physiol. Behav., vol. 176, no. 1, pp. 139–148, 2020, doi: 10.1146/annurev-pharmtox-010919-023324.Big.

[34] Krishnaveni, N., Nagaprasad, S., Khari, M., & Devi, V. A. (2018). Improved Data integrity and Storage security in Cloud Computing. *International Journal of Pure and Applied Mathematics*, *119*(15), 2889-2897.

6

Cloud Computing Environment for Big Data Analytics

Sushma Malik[1], Ankita[1], Madhu[1], and Anamika Rana[2]

[1]Institute of Innovation in Technology and Management, India
[2]Maharaja Surajmal Institute, India

Email: sushmamalikaiitm@gmail.com; ankkita21@gmail.com;
myself_madhu26@yahoo.com; anamica.rana@gmail.com

Abstract

Nowadays although data is growing more rapidly, it is more crucial to understand how to handle it and extract hidden facts from it. The speedy development of digital technologies has led to a significant increase in the volume of digital data. As a result, a lot of data is generated by various sources, such as social media, smartphones, sensors, etc. It takes time and effort to effectively manage and evaluate big data. Decision-makers can uncover significant information from big data after processing it through in-depth analysis and effective data processing. Different methods and approaches of big data analytics can be utilized to extract worthwhile insights from such diverse and quickly expanding datasets, which may improve strategic planning and decision-making. Cloud computing is a powerful model that can handle and store enormous amounts of data. Cloud computing provides on-demand, flexible, accessible, and reasonably valued resources by doing away with the need to maintain hardware, software, as well as space on servers. Both technologies are constantly changing and work well together. Scalability, agility, and elastic on-demand data availability are advantages of integrating them. Clusters of servers are required in a big data environment to maintain the tools necessary to process enormous volumes of data quickly and in a variety of formats. Cloud computing economically provides this kind of service by giving users access to a cluster of computers, storage, and networking

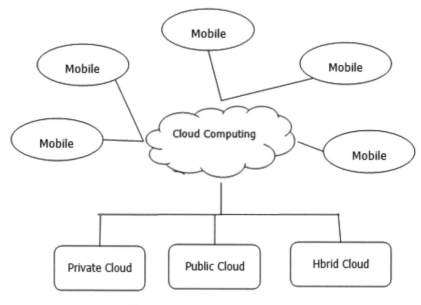

Figure 6.1   Cloud computing [4].

resources that can scale up or down as needed. With the use of cloud computing, several users can access and update their data on a single server without
having to purchase additional apps. This chapter examines the function of
big data within the context of cloud computing. The definition and classification of big data with its tools and challenges are included in this chapter
with cloud computing technology definition and its types. The architecture of
cloud computing in big data technology is also elaborated on its advantages.

6.1 Introduction to Cloud Computing

No doubt cloud computing is one of the most popular technologies in the
world of information technology throughout the world. Before exploring the
term in-depth, we must know what this term is. Cloud is nothing but a virtual
space that delivers remotely on the Internet, the myriad services including
server, software, network, and storage, and there is much more in the bag
to mention as shown in Figure 6.1. The organization or customer needs to
pay what he requires. This not only provides you with the best services in a
faster and optimized manner but also lowers your operating cost. The companies that provide cloud services allow users to store documentation, files,
software, and configurations on remote servers which enable users to quickly

Figure 6.2   Advantages of cloud computing.

access information from any part of the world as the information is omnipresent. Cloud takes away all the heavy files and applications from the hard drive to cyberspace on the virtual server thereby reducing the load of the client machine which also makes the client system faster [1–3].

Thus, in today's world, cloud computing is the most popular option for organizations and other users to increase productivity, speed, efficiency, and performance. Additionally, the cloud does not need much effort to start and maintain. As it is infinite in size, so no one needs to worry about running out of space. To grab all the benefits of cloud computing one thing you need is just an Internet connection [4, 5].

6.2 Advantages of Cloud Computing

As of now, we all know what cloud computing is. Various benefits have made the technology popular and acceptable worldwide. This technology has become an indispensable part of a large number of organizations. Earlier as the technology was unexplored, people were not aware of its benefits but now when cloud computing has evolved with time, people are more motivated and willing to implement the technology in small to large organizations [6, 7].

Some of the benefits are shown in Figure 6.2.

6.2.1

6.2.1.1 Economy
Cloud computing does not require the purchase of exorbitant hardware and software devices for its setup. There is no need to arrange a rack of servers, electricity for power and cooling, and professionals to manage the infrastructure.

6.2.1.2 Speed
In the IT sector where the computational time has reduced to nanoseconds, speed matters. All information must be available in just a fraction of a second. Cloud keeps this aspect into consideration and delivers the information in no time. Businessmen and entrepreneurs need to devote more time to analyzing the data and decision-making rather than waiting for content to get downloaded. This in turn improves throughput.

6.2.1.3 Customization
The more the number of minds, the more the ideas as so are the preferences. Every client has different requirements and expectations from the cloud. The cloud services can be customized like more or less computing power, the bandwidth required, and the storage capacity needed, and accordingly, the client needs to pay charges for what is demanded.

6.2.1.4 Productivity
Productivity depends upon how much effort you put into the specified time constraint. As cloud computing technology does not require any kind of racking and stacking, software patch-up, or hardware setup. So the employees can concentrate and devote time to other important tasks instead of wasting time on less important activities.

6.2.1.5 Reliability
Entrepreneurs can have complete faith and rely on the technology of the cloud for data recovery. In this technology, the data is mirrored and stored at multiple locations for data backup. The client-provider network uses advanced methods for data recovery and disaster management and thus keeps the data available whatever the situation may be.

6.2.1.6 Security
Most cloud providers offer a wide set of policies, technologies, and controls to strengthen the security of your data. The various security mechanisms help in protecting your data, applications, and infrastructure from threats.

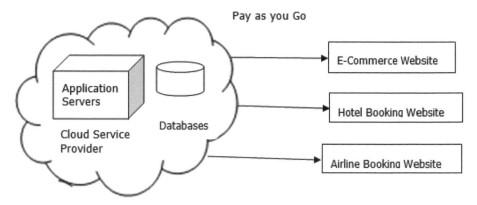

Figure 6.3 Public cloud [11].

6.2.2 Type of clouds – the ways to deploy

Depending upon how to deploy the cloud services can be categorized into three different categories: a public cloud, private cloud, and hybrid cloud. In this section, we will study each type in detail [6, 8, 9].

6.2.2.1 Public cloud

As the name suggests, public clouds are owned and operated by third-party cloud service providers. The third party provides their services like server, memory, and computational power over the Internet according to the client's needs. All the types of hardware requirements and software requirements along with bandwidth and infrastructure needs are taken care of by a third party only. The services provided by third parties may be free (for a particular time period) or paid. The client needs to access the services and use the cloud as and when required as shown in Figure 6.3.

The customer is required to pay for the customized services according to the features, services, and resources the organization has opted for.

For example, Microsoft Azure, Amazon Web Services (AWS), Google cloud, IBM cloud, and Oracle cloud are public clouds [5].

The main benefit of the public cloud is that they save organizations from investing in exorbitant devices and facilities thereby providing the services on a demand basis. Apart from providing the facilities, the third party provides maintenance and technical support. In short, the third party is responsible for the complete package, starting from procurement to implementation to maintenance and support [10].

As the facility is on the Internet so any employee of the company can access the services at any time and from any location as long as they have Internet connectivity.

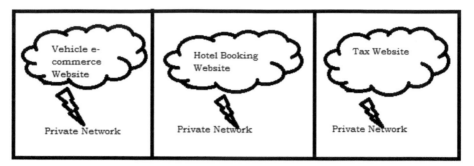

Figure 6.4   Private cloud [11].

Many organizations are attracted by the benefits of the public cloud includ-ing elasticity, flexibility, and adjustable workload demands, and are shifting toward this type of cloud. Additionally, efficiency has increased, and also there is minimal wastage of resources, which also lures businessmen and entrepreneurs.

6.2.2.2 Private cloud

Private means something owned by oneself. It is the cloud that is dedicated to a single organization. In this type of cloud, all the hardware, software, memory, bandwidth, and everything is taken care of by a single organization. A private cloud is physically located in the company's on-premise data center (Figure 6.4). In case of scarcity of space, a private cloud can also be hosted on rented infrastructure housed in an offsite data center.

It provides many benefits like elasticity, scalability, and ease of service to deliver desired resources.

Some companies can set up on their own while others may seek techni-cal support from third parties. The data available on a private cloud is to be used by people of the organization rather than the general public.

A private cloud is also known by the name of a corporate cloud or inter-nal cloud.

Many companies prefer the private cloud because they deal with con-fidential information/documents, financial data, intellectual property, or other sensitive data.

To make the data available only to a selected number of users, a high level of security is required by internal hosting and company firewalls.

6.2.2.3 Hybrid cloud

In the above section, we have discussed the pros and cons of private and pub-lic clouds. There is another approach in which private cloud and public cloud can be bundled together. Combining both types of clouds, greater flexibility,

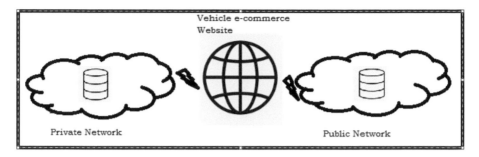

Figure 6.5   Hybrid cloud [11].

more deployment alternatives, and optimization of existing techniques can be achieved in a business as shown in Figure 6.5.

Like a public cloud, it gives access to other users who wish to access data from the cloud, and at the same time, it takes care of proper restrictions which should be imposed on users to access information from the cloud.

Therefore, we can say that the hybrid cloud dilutes the concept of isolation but at the same time provides boundaries. In the case of a hybrid cloud, one can integrate or aggregate it with another cloud to enhance its capacity and capability.

The concept of a hybrid cloud can be understood with the help of some use cases. For example, an organization can store sensitive and private data on a private cloud but share it with some authorized selected users or clients. In this use case, we are using the concept of private and public clouds at the same time. In this case, the organization needs to pay special attention to access mechanisms and the security of data.

In another example, an organization, in case of limited space on a private cloud can use public cloud computing resources. In this technique, clouds need to scale themselves to other public clouds.

Here, public clouds come into the picture on demand for computing capacity. The main advantage here is that the capacity of the cloud can be increased as per demand indeed by paying extra throughout the year.

6.2.3 Service models of cloud

In 2011, the National Institute of Standards and Technology divided cloud computing into three service models, namely [12]:

- SaaS (stands for software as a service),

- IaaS (stands for infrastructure as a service),

- and PaaS (stands for platform as a service).

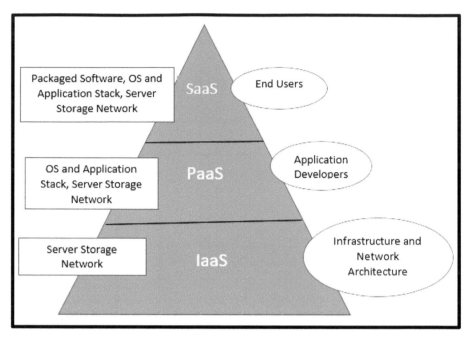

Figure 6.6   Cloud service models [13].

These three service models provided services in their field of work and provided scope for researchers, programmers, and technicians to experiment and explore. All these types of services can be depicted in the form of a stack as highlighted in Figure 6.6. This is sometimes called the cloud computing stack. This stack model offers increasing abstraction; they are thus often viewed as layers in a stack (Figure 6.7): infrastructure-, platform-, and software as a service, but these need not be related [5].

6.2.3.1 Software as a service (SaaS)
Software as a service is a mechanism to deliver applications and software over the Internet. Instead of installing the software on the client machine, one can install and maintain it on the cloud. Due to this, there is no need for a user to overburden the computer's memory [12, 15].

6.2.3.2 Infrastructure as a service (IaaS)
This is one of the basic categories of cloud computing services. In this type of service, the organization takes all the IT infrastructure like servers, network, location, storage, and security on rent from a third-party service provider and pays the charges according to the services opted for. The cost of setup

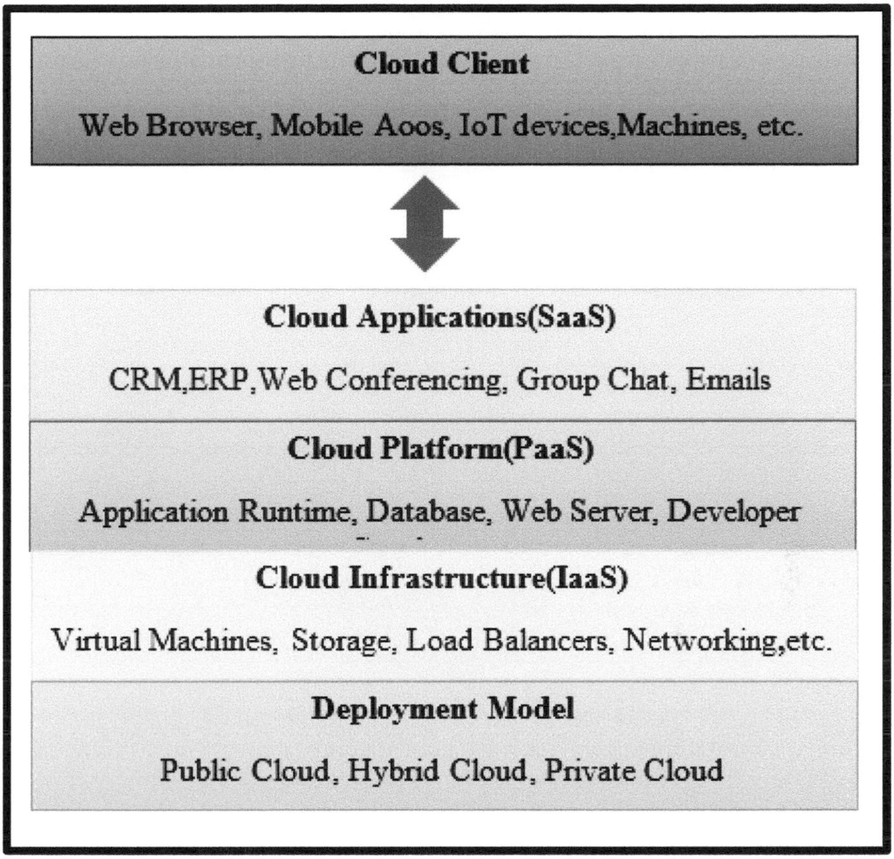

Cloud Client

Web Browser, Mobile Aoos, IoT devices,Machines, etc.

Cloud Applications(SaaS)

CRM,ERP,Web Conferencing, Group Chat, Emails

Cloud Platform(PaaS)

Application Runtime, Database, Web Server, Developer

Cloud Infrastructure(IaaS)

Virtual Machines, Storage, Load Balancers, Networking,etc.

Deployment Model

Public Cloud, Hybrid Cloud, Private Cloud

Figure 6.7   Cloud deployment model [14].

depends upon the number of services opted for and consumed. In this type of service, the customer does not have any control over cloud infrastructure, rather it has control over the operating system, storage, and applications along with some limited control over firewalls and security aspects [12, 15].

6.2.3.3 Platform as a service (PaaS)

Platform as a service, as the name suggests, refers to cloud computing services that supply an environment for developing an application or software, testing, delivering, and managing software applications. The capabilities provided to customers include programming languages, libraries, services, and tools. In short, the provider offers a toolkit and standards for development.

PaaS makes it easier and more convenient for developers to design web applications or mobile applications in a faster manner. The client need

Figure 6.8   Characteristics of cloud computing.

not worry about the underlying infrastructure of servers, databases, storage, etc. [12, 15].

Figure 6.8 highlights some of the key properties of cloud computing [16, 17].

- **Elasticity**: Different companies may have different types of requirements depending upon the capital and other restrictions because some might need infrastructure, some might need only memory, and others may require computational powers or such requirements in any combination. The companies can scale up and down the services required.

- **Flexi payment**: Customers and companies can select the service according to their requirements and pay the required amount accordingly. There is no fixed amount or package that the customer has to stick to. This gives a kind of flexibility or elasticity to customers.

- **Resistance**: The communications service providers often provide resistance facilities to ensure that bandwidth, computational power, and memory are available all the time and users can access the facilities across multiple locations on the globe without any technical glitch or failure.

- **Easy migration**: As we know that organizations are highly dynamic so there may be a need to move from one cloud to another or merge more than one cloud to gain the advantage of a hybrid cloud. Cloud computing technology provides a seamless migration and merging of clouds whenever required.

- **Resource and tenancy pool**: Cloud computing also allows multiple customers to share the same physical infrastructure or same memory or any other service while maintaining privacy and security. At this point in time, cloud providers need to be extra cautious while delivering the service. The IT behind the cloud has to provide the required service with proper security mechanisms.

- **Recovery**: If any end user is concerned about data recovery, then the cloud has a solution for it too. In case of any damage to the laptop, mobile phone, or desktop the data remains safe, secure, and available. The highly secured algorithms implemented on the cloud always keep the data available to users and ensure that any damage to devices does not cause any damage to data and thus does not hamper regular work.

6.2.4 Disadvantages of cloud computing

As every coin has two sides, in the same manner, every technology has pros and cons. Although, the number of advantages outnumbers and outweighs the disadvantages still there are some limitations to mention. Some of the limitations are mentioned in Figure 6.9 [18, 19]:

- **Resistance of users**: Despite high-level security mechanisms, some users fear data breaches, hacking of APIs, and loss of authentication and credential details. due to some reasons, they refrain from implementing the cloud.

- **Unpredictable cost**: As mentioned in the above sections, customers can pay for the cloud according to the facilities and services that they opt for. This, sometimes, may lead to unpredictability in terms of investment forecasting and investment calculation. Additionally, any change made in services during the financial year may lead to additional and unpredictable costs.

Figure 6.9   Disadvantages of cloud computing.

- **Lack of technical staff**: With the fast-growing technology of the cloud, there is a scarcity of highly technical, qualified technical staff to implement and maintain the cloud systems installed at different locations. There is a demand for employees who are well-trained in the tools and applications required for cloud computing.

- **Conflicting industry laws**: At the time of transferring the data to the cloud the data must be provided to a third party, at the same time the company must be aware of how and where the data is hosted. This sometimes leads to conflict in regulatory compliance.

- **Challenging task**: It becomes a challenging, hectic, and difficult task for the IT department to design architecture, and build and maintain a hybrid cloud. In addition to this, cloud migration is also a cumbersome task and sometimes takes a lot of time which usually leads to over budget.

6.2.5 Cloud computing use cases

Cloud computing has entered into almost every sector and in the majority of organizations as people at large have identified and trusted the features

BIG DATA

Figure 6.10   Big data [22].

and services offered by cloud computing. Some of the use cases are as follows:

- **Microdoft365 and Google Docs**: Users can use Microsoft 365 and Google Docs are extensively used by a large number of users because one can create and access documents, worksheets, and presentations stored in the cloud at any time and from anywhere.

- **WhatsApp, email access**: Users can take advantage of the cloud ability to remotely access email notifications, content, and WhatsApp messages through their desktop or laptop, or any other device.

- **Virtual meetings**: Zoom is one of the tools which are based on a cloud-based platform to conduct online meetings, webinars, and conferences. Apart from attending virtual meetings, one can record the session, send reactions, chat with participants, access the list of attendees, and much more.

6.3 Definition of Big Data

At present, Data is becoming more precious, however, the issue remains the same as to how to manage data and look for hidden facts in this huge data. Big data is a broad term used for vast and complex datascts that are too big, growing rapidly, and moreover tricky to manage by traditional tools & methodology. Big data is not restricted to any domain. It may be gathered from enormous sources; be it smartphones, sensors, auditory and video inputs, and the big dataset giant social media as shown in Figure 6.10. All these sources are escalating the amount and diversity of data. Big data has the prospective to give priceless facts afterward processing that can be revealed in the course of in-depth study and competent processing of data by verdict announcers. To take out important patterns from such diverse and fast-rising datasets, a variety of tools and methodologies of big data analytics can be incorporated that may escort to enhanced decision-making and deliberate forecasting. Big data skill is basically dependent on parallel processing, in-database implementation, perfect management of storage space and assorted workload administration [20, 21].

At the commencement of the twenty-first century, the rising amount of data offered an apparently inexplicable crisis; storage and CPU (central processing unit) technologies were snowed under by the terabytes of data being produced. Luckily, Moore's law came to the salvage and assisted to make storage and CPUs superior, quicker, smarter, and less expensive. Today, big data is no longer a scientific crisis: it has turned into a viable gain. Enterprises are mounting and utilizing big data means for discovering their data channel, finding out novel approaches that might lend them a hand to expand improved relations with their clients, classifying fresh regions of business openings, and by administering their supply chains healthier, all in a gradually more viable business atmosphere. In small, big data can give out to get better services, products, and processes, all the way through its impact on decision-making and data processing.

6.3.1 Characteristics of big data

Big data has led five V's as shown in Figure 6.11 [23, 24]:

1. **Volume**: refers to enormous amounts of produced and stored data that are measured in ZB or YB rather than TB. The "size" of the data refers to its usefulness and whether or not it can be classified as "big data."

2. **Variety**: belongs to several types of data, such as structured, semi-structured, unstructured, and raw data, that are used. To manage all of the available raw data, several strategies and methodologies are needed because different data types are available.

3. **Velocity**: since ages, it is a synonym for "speed" and in the big data context, it refers to the speed with which new data is generated and the speed at which it is moving in the market.

4. **Variability**: relates to the existence of "discrepancy" or irregularity during data analysis or at the time of the foundation of summaries. For a data analyst, it can be challenging to grasp it adequately.

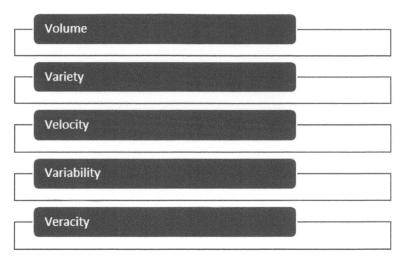

Figure 6.11   Characteristics of big data.

5. **Veracity**: the "vagueness" that results from inconsistency and irregularity in the data.

6.3.2 Cloud-based big data management tools

To handle the massive amount of planned and unarranged data, big data creates tremendous problems. So that everyone may make use of technology, cloud computing provides scalable methods to manage such a big volume of data in a cloud environment. Understanding the tools and services they offer is essential for effectively integrating and managing big data in a cloud context. Cloud-based Hadoop and NoSQL database platforms are being proposed by some vendors, including Google, IBM, and Amazon Web Service (AWS), to support big data claims.

These are not the only dealers who propose cloud-based database platforms; many cloud dealers do provide their big data services like AWS's Elastic MapReduce, Google's BigQuery, etc. The majority of the cloud service supplier proposes Hadoop to support that balance of the regular requirement of clients routinely for data processing [25].

Hadoop

Java-based distributed storage and processing applications on very big datasets are supported by the open-source Hadoop framework. The Hadoop platform has top-tier declarative languages for writing pipelines for data analysis and query scripting. Hadoop has several components, but the distributed file system (HDFS) and MapReduce are the most commonly advised ones for big data. The additional parts offer balancing services and a higher level of generalization. For instance, companies like Google, Twitter, and Facebook gained attention for handling massive amounts of data in the cloud by harvesting data using comments made on social media websites. The information obtained enhanced their products and reduced costs.

MapReduce

The essential part of the Hadoop framework is the MapReduce system, which is used to analyze and produce large datasets in a group using a scattered or equivalent method. It is a programming model that divides the task into various self-governing nodes to practice handling enormous amounts of data. A MapReduce program combines two methods: the Map technique, which involves gathering, examining, and categorizing datasets, and the reduce method, which involves creating summaries and producing final results. MapReduce system assembles dispersed servers, handles all connections, and similar data transfers, and also offers redundancy and error acceptance.

Hadoop distributed file system (HDFS)

Large data files, often in the gigabyte to terabyte range, which cannot fit on a single system are stored using HDFS. For the Hadoop framework, a distributed, scalable, and controlled file system known as HDFS was created in Java. To facilitate parallel processing, it maintains reliability by duplicating data over various hosts. To do this, it divides a file into blocks that will accumulate across multiple workstations. The HDFS group consists of a single name node, several data nodes, and a master-slave association.

Cassandra and HBase

Open-source, distributed, nonrelational DBMSs written in Java called Cassandra and HBase provide structured data storage for large tables and run on top of HDFS. It is a columnar data paradigm with in-memory operations and compression capabilities.

Hive

It is a storehouse infrastructure by Facebook, which offers data consolidation, ad hoc querying, and study. It offers a query language i.e. HiveQL to create dominant queries and produce the outcome in actual time.

Zookeeper

It upholds a raised show synchronization administration for a different application that can load up setup data and have an ace treatment hub.

6.3.3 Big data lends help in various forms in a business

6.3.3.1 Assist in employment

Big data helps the spotters for getting to the apparatuses to put the up-and-comers as soon as conceivable to contend. With the assistance of huge information, the enrollment specialists are currently ready to accelerate and robotize the arrangement interaction.

The big data enlistment stage can give an inner data set in 360-degree perspective on an up-and-comer, which can certify their training abilities or whatever else an individual has.

6.3.3.2 Log examination

It is one of the significant perspectives in business as log the executives and examination devices have been around quite a while in the past. The log information can turn into prevention when we want to store, cycle, and present practically and productively.

Many apparatuses are accessible on the lookout and can give offices to process and investigate huge log information without unloading it into the social data set. This lost information can be recovered with the assistance of SQL questions.

6.3.3.3 Advantages of e-publishing

Big data extends the endorser base and lift the main concern and contributes extraordinary Website design enhancement work to make the distributing webpage accessible. Big data think of groundbreaking thoughts for handling and investigating both substance information and client information.

With the assistance of a strong web index, the records stay clean. What is more, with the assistance of AI prescient investigation, the distributor gives better content to its purposes.

6.3.3.4 E-commerce

E-commerce enterprises are benefiting greatly from strong search and big data analytics.

Additionally, it combines predictive analysis and machine learning to infer the user's preferences from log data, improving the search experience for both laptops and mobile devices.

6.3.3.5 Search engines

Big data is versatile and strong and can deal with a lot of information. With the assistance of this element, organizations empower to investigate the information.

We can recover this information from clients and proposes better proposals with the assistance of AI and prescient examination, these suggestions are getting better step by step.

6.3.3.6 Decline in scam

Today, information is produced in enormous numbers, and in exchange; we require more creative and successful thought for disposing of the cheats. Taking into account conventional misrepresentation discovery models the specialists examining extortion worked with examination to run complex SQL questions.

Ordinarily, this cycle caused trust in cases, and alongside it, there was an enormous misfortune for the business. So huge information gives abilities front identification and gives programmed alarms which perceive the issue and dispense with it soon.

6.3.4 Challenges

Challenges are not restricted to one domain. Rather it is distributed in Ethical, privacy, and security challenges [26].

6.3.4.1 Ethical challenges

The significance of morals is exceptionally huge in enormous information examination, particularly about the profiling of clients through web-based entertainment investigation. Simultaneously, individuals are progressively becoming mindful of how their information is being utilized, even as certain people and firms are exceptionally quick to exchange individual information for business gains. Subsequently, explicit regulations and guidelines should be laid out to protect the individual and delicate data of clients, as no particular regulations address these obtrusive and meddling advancements.

6.3.4.2 Privacy challenges

As innovation advances, more brilliant gadgets and machines are given the ability to interface with networks and make accommodations for clients. Albeit the advantages of these gadgets are critical, the potential disadvantages can be made when ill-advised execution conventions are utilized to convey data among networks. Thus, estimating the security of these connections is significant.

Sending information could give delicate insights concerning the proprietor's well-being or religion, similarly as the flawed arrangement of information classification and respectability could likewise impact clients' protection as noxious gatherings can break individual data with next to no approval or assent. There is a developing worry among clients over the absence of security insurance for their information by associations' information assortment strategies, particularly concerning the utilization of following innovations, like treats and GPS trackers.

6.3.4.3 Security challenges

The security challenges presented by the flood of enormous information innovation and the web availability of savvy gadgets appear to be unabated. It is anticipated that toward the finish of 2020, there will associate with 20 billion associated gadgets.

Likewise, the greater part of this information comes in various arrangements, for example, organized and unstructured. Albeit the principal concern is a large part of the unstructured information as message information, messages, texts, sound and video pictures, street traffic data, and mixed media objects, which are delicate and may contain individually recognizable data (PII) and protected innovation (IP).

6.4 Cloud Computing Versus Big Data

Cloud computing and big data (Table 6.1) are the two main technologies in the field of IT and they have played an essential role. But they have also some differences, which are highlighted here [27].

6.4.1 The usefulness of integration of big data and cloud computing

Cloud computing and big data are crucial to a business. They are the main focus of attention in the company because this technology contributes to growth and increased productivity. For company expansion, decision-making, and other purposes, big data, and cloud computing are used. A technique called cloud computing offers virtual services. With the aid of the Internet, we may access these services, which are secure and trustworthy. Small businesses benefit greatly from it because it offers storage space at low costs and

Table 6.1   Cloud computing versus big data.

A comparative point	Cloud computing	Big data
It is what? Focus	Computing model ensuring service accessibility for all	large-scale data collections Fix the technology issue of handling enormous data sets
Best explained by	In cloud computing, services are delivered over a network, mostly the Internet. Software, a platform, or an IT infrastructure are all examples of services.	Velocity, Volume, and Variety are the "3 Vs." The data set of interest must demonstrate any or all of the aforementioned Vs for it to be considered "big data."
When should move?	Moving to the cloud while preserving centralized access when quick deployment or scaling of IT applications or infrastructure is required. With using cloud computing, the focus is still on business, unlike when maintaining IT operations on-premise.	When dealing with a massive amount of data, standard approaches and frameworks become inadequate, which is when big data engineering comes into play. Petabyte-scale data analysis necessitates parallel processing and a distributed infrastructure.
When to not move?	On the other hand, some people may decide against moving to the cloud. Keep items off the cloud if the application handles extremely sensitive data, demands stringent compliance, or does not follow cloud architecture.	In contrast, moving to the cloud may not be desired in some circumstances. Keep things away from the cloud if the application handles extremely sensitive data, demands stringent compliance, or does not follow cloud architecture.
Benefits	Low up-front costs, disaster-safe installation, centralized platform, and low ongoing costs.	High scalability (scales out forever), cost-effective, parallelism, reliable ecosystem.
Common roles	The administrator of cloud 1. **Resources**: An individual or group in charge of managing the cloud. 2. **Cloud service provider**: The person or company that owns the cloud platform and offers services in the form of resources, applications, or infrastructure	1. **Developers of big data**: They create software to absorb, process, or purify data. They also built-up systems for scheduling and delta capture. 2. **Big data administrators**: These professionals configure servers, set up applications, and manage logical or physical resources.

	3. **Cloud consumer**: A cloud user may be a developer or an office employee within a business. 4. **A cloud service broker** is a third party that acts as a conduit between customers and service providers. The services they offer are intermediaries. 5. **Cloud auditor**: The person who advises Customers on security or potential vulnerabilities	3. **Big data analysts**: These people are in charge of looking through the data to uncover intriguing patterns and potential future trends. 4. **Data scientist**: Essentially an analyst with coding and statistics expertise. This individual works on data mining, predictive modeling, and data visualization projects involving big data systems. 5. **Big data architect**: This person is in charge of deploying end-to-end solutions.
Buzz words	When service providers give consumers access to physical resources like memory, disc space, servers, and networking, this is known as infrastructure as a service (IaaS). The consumer installs applications on top of these services and is free to use them in any way she pleases. PaaS: An operating system, RDBMS system, server, or programming environment are all examples of platforms. Platform as a Service is the delivery model used for all of these platforms. In the SaaS model, the consumer directly uses the application or software and is not concerned with the infrastructure or platform that supports it.	Hadoop is a popular term right now. It is an ecosystem of different parts that each performs a specific function and are combined to implement a big data solution. Doug cutting took inspiration for the project's name from his son's elephant-shaped toy. High throughput access is made possible via the file system known as HDFS (Hadoop distributed file system). It is a distributed file system built on Java that utilizes numerous computers. Massively parallel programs that process vast amounts of HDFS-stored data can be created using the MapReduce framework. At its most basic, MapReduce consists of two operations: Map, which turns data into Key-Value pairs, and reduces, which aggregates the data.
Vendors/solutions providers	Google, Amazon, Microsoft, IBM, Dell, Apple	Cloudera, MapR, HortonWorks, Apache

effectively leverages their data for research. The combination of big data and cloud computing is ideal for corporate expansion [28].

6.4.2 The architecture of cloud in big data

Cloud computing and big data have many similarities. Big data primarily focuses on value extraction, whereas cloud computing is more concerned with scalable, elastic, on-demand, and pay-per-use self-service models. Big data demands enormous on-demand processing power and dispersed storage, yet cloud computing effortlessly offers the elastic on-demand integrated computing resources, necessary storage, and processing power needed for big data analysis [29]. Along with distributed processing, cloud computing provides virtual machines for scalability and extension to meet the demands of exponential data growth. As a result, analytical platforms have grown in number to meet customer demands, particularly those of big data-driven businesses, for contextually evaluated data to be provided from all the information that has been stored [30].

A brand-new, excellent method for offering IT-related services is the cloud computing environment. There are numerous providers and user terminals in the cloud computing ecosystem. It covers various types of hardware and software, as well as pay-per-use or subscription-based services provided both online and in real time. Big data tools are used to gather data, which is then saved and processed in the cloud. The cloud offers resources and services on demand for continuous data management. The most popular big data analytics models are platform services like (PaaS), infrastructure services like (IaaS), and software services like (SaaS) (IaaS) [24].

The input, processing, and output (IPE) paradigm is used to describe the link between big data and cloud computing (Figure 6.12). The input is large data in a structured, unstructured, or semi-structured format that has been gathered from a variety of data sources, including cellular and other smart devices. This massive amount of data is cleaned before being placed in Hadoop or another data repository. To provide services, the saved data is processed again using cloud computing tools and techniques. The tasks necessary to transform input data are all included in the processing phases. The value acquired during data processing for analysis and visualization is represented by the output [31].

One of the elements in common between cloud computing and big data is the Internet of Things (IoT). IoT device data generation is huge, and it requires real-time analysis. Cloud service providers keep data in data stores and enable data transmission via the Internet or lease lines. The huge data that has been stored in the cloud is then filtered and analyzed using cloud

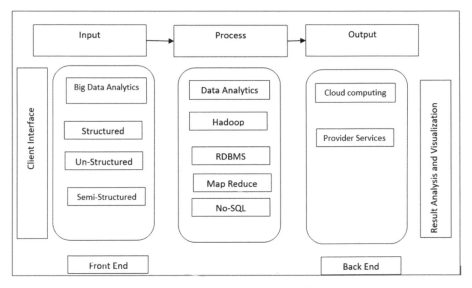

Figure 6.12   The architecture of cloud in big data.

computing techniques and technologies. It offers a mechanism for the data to travel, be kept, and be examined. Big data and IoT have a shared platform thanks to cloud computing. Big data, IoT, and cloud computing all work well together [32].

6.4.3 Advantages of using cloud for big data analytics

Despite the evidence supporting the cost-benefits, competitive advantages, and business efficiency of cloud computing, a sizable portion of the corporate world still operates without it. However, as data volume increases, it becomes more and more challenging for individuals and organizations to keep all of their vital data, programs, and systems up and running on internal computer servers. One workable method for solving this issue is to use cloud computing. Resource pooling is used in cloud computing environments, which are built for general-purpose applications, to provide flexibility on demand. The cloud computing infrastructure thus seems to be the most appropriate for big data. In this part, we have discussed a variety of benefits of using big data in the cloud (Figure 6.13). big data and cloud computing together are just amazing! [28, 29, 33].

- **Make analysis better**: Big data analysis has become better thanks to the development of cloud technologies, producing superior conclusions. For a deep view of data, several cloud-based storage alternatives

Figure 6.13   Advantages of using cloud for big data analytics.

come with cloud analytics built in. Users may quickly implement tracking systems and design custom reports to examine data across the entire enterprise with data in the cloud. Users can increase productivity and develop action plans based on the results to meet the needs of the organization.

- **Offer a robust infrastructure**: Workload management is made simple by the flexible infrastructure provided by cloud computing, which may grow in response to demand.

- **Reduce cost**: Both big data and cloud computing help organizations by reducing ownership expenses. Customers can process big data in the cloud without spending money on expensive big data infrastructure. Fewer expenses and more returns result from the combination of these traits.

- **Data security and protection**: Data security and privacy are two essential issues to take into consideration while working with company data. Furthermore, when your application is on a cloud platform, security becomes a crucial factor due to the open environment and limited user permissions. System integrators now provide a Private cloud Solution that is more elastic and scalable. Scalable Distributed Processing is also used. Additionally, networks are used to transfer encrypted data, increasing security. When data is encrypted, hackers and anyone who should not have access to it have a harder time acquiring it.

- **Provisioning quickly**: Cloud-based platforms enable businesses to deploy massive databases as needed, unlike traditional big data analytics platforms that need a time-consuming and expensive process to acquire the required hardware and software. It takes only a few minutes to spin up thousands of servers.

- **Higher availability and efficiency**: Cloud-based infrastructures use high availability architecture, in contrast to conventional warehousing platforms, to reduce downtime. Additionally, a virtualized system shares pooled resources and distribute workload across multiple applications, leading to increased efficiency.

- **Relevant real-time analysis**: To speed up the data analysis process, cloud-based platforms make use of an elastic collection of resources, providing result sets quickly enough to offer business-relevant analysis and insight – in real-time. Companies wanting to provide their clients with more individualized and focused experiences can greatly benefit from this.

6.5 Conclusion

The amount of data available now is enormous and keeps growing daily. The types of data being produced are likewise becoming more varied. The proliferation of mobile smartphones and other gadget sensors connected to the Internet has increased the rate of data collection and expansion. Businesses in many sectors can make use of these data opportunities to acquire current business insights. It has been possible for some time to use cloud services to store, process, and analyze data; this has altered the context of information technology and made the on-demand service model's promises a reality. We put forth a big data taxonomy, a conceptual understanding of big data, and a cloud services model. Several representative big data cloud platforms were compared to this paradigm. We have provided a thorough analysis of how big data concerns can be effectively addressed by cloud computing and its related distributed computing technologies in this chapter.

References

[1] B. Berisha, E. Mëziu, and I. Shabani, "Big Data analytics in Cloud computing: an overview," J. Cloud Comput., vol. 11, no. 1, 2022, doi: 10.1186/s13677-022-00301-w.

[2] T. S. Harsha, "Big Data Analytics inCloud computing Environment," vol. 8, no. 8, pp. 159–179, 2017.

[3] A. Sunyaev, "Cloud computing," Springer, pp. 195–236, 2020, doi: https://doi.org/10.1007/978-3-030-34957-8_7 195.

[4] C. Baird, "A Primer on Cloud computing. Cloud computing is defined as: | by Colin Baird | Medium," 2019. https://medium.com/@colin-baird_51123/a-primer-on-cloud-computing-9a34e90303c8 (accessed Sep. 24, 2022).

[5] N. Reckmann, "How Cloud computing Can Benefit Your Small Business," 2022. https://www.businessnewsdaily.com/4427-cloud-computing-small-business.html (accessed Sep. 24, 2022).

[6] M. Ali, S. U. Khan, and A. V. Vasilakos, "Security in cloud computing: Opportunities and challenges," Inf. Sci. (Ny)., vol. 305, no. January, pp. 357–383, 2015, doi: 10.1016/j.ins.2015.01.025.

[7] D. Puthal, B. P. S. Sahoo, S. Mishra, and S. Swain, "Cloud computing features, issues, and challenges: A big picture," Proc. - 1st Int. Conf. Comput. Intell. Networks, CINE 2015, pp. 116–123, 2015, doi: 10.1109/CINE.2015.31.

[8] I. Odun-Ayo, M. Ananya, F. Agono, and R. Goddy-Worlu, "Cloud computing Architecture: A Critical Analysis," Proc. 2018 18th Int. Conf. Comput. Sci. Its Appl. ICCSA 2018, pp. 1–7, 2018, doi: 10.1109/ICCSA.2018.8439638.

[9] J. Clark, "Cloud Service Models SaaS vs IaaS vs PaaS: Which One You Should Choose and Why? | Branex - International," 2020. https://www.branex.com/blog/cloud-service-models-saas-vs-iaas-vs-paas/ (accessed Sep. 24, 2022).

[10] W. Kim, "Cloud computing: Today and Tomorrow," J. Object Technol., vol. 8, no. 1, pp. 65–72, 2009, doi: 10.5381/jot.2009.8.1.c4.

[11] A. Kumar, "Cloud Deployment Models | Public, Private and Hybrid," 2020. https://k21academy.com/cloud-blogs/cloud-computing-deployment-models/ (accessed Sep. 24, 2022).

[12] M. Al Morsy, J. Grundy, and I. Müller, "An Analysis of the Cloud computing Security Problem Mohamed," Apsec, pp. 1–6, 2010.

[13] A. Fu, "7 Different Types of Cloud computing Structures | UniPrint.net," 2022. https://www.uniprint.net/en/7-types-cloud-computing-structures/ (accessed Sep. 24, 2022).

[14] D. Rountree and I. Castrillo, "Cloud Deployment Models," Basics Cloud Comput., pp. 35–47, 2014, doi: 10.1016/B978-0-12-405932-0.00003-7.

[15] M. Birje, P. Challagidad, M. T. Tapale, and R. H. Goudar, "Security Issues and Countermeasures in Cloud computing Cloud computing review: concepts , technology , challenges and security," no. June 2020, 2015.

[16] M. I. Malik, "Cloud computing-Technologies," Int. J. Adv. Res. Comput. Sci., vol. 9, no. 2, pp. 379–384, 2018, doi: 10.26483/ijarcs.v9i2.5760.

[17] C. Stergiou, K. E. Psannis, B. G. Kim, and B. Gupta, "Secure integration of IoT and Cloud computing," Futur. Gener. Comput. Syst., vol. 78, pp. 964–975, 2018, doi: 10.1016/j.future.2016.11.031.

[18] S. Namasudra, "Cloud Computing: A New Era," J. Fundam. Appl. Sci., vol. 4, no. 4, pp. 527–534, 2018.

[19] N. Subramanian and A. Jeyaraj, "Recent security challenges in cloud computing," Comput. Electr. Eng., vol. 71, no. July 2017, pp. 28–42, 2018, doi: 10.1016/j.compeleceng.2018.06.006.

[20] I. Yaqoob et al., "Big Data: From beginning to future," Int. J. Inf. Manage., vol. 36, no. 6, pp. 1231–1247, 2016, doi: 10.1016/j.ijinfomgt.2016.07.009.

[21] C. W. Tsai, C. F. Lai, H. C. Chao, and A. V. Vasilakos, "Big Data analytics: a survey," J. Big Data, vol. 2, no. 1, pp. 1–32, 2015, doi: 10.1186/s40537-015-0030-3.

[22] M. Galetto, "NGDATA | Learn Big Data Analytics: 51 Expert Tips," 2016. https://www.ngdata.com/mastering-big-data-analytics/ (accessed Sep. 24, 2022).

[23] Y. Zhang, J. Ren, J. Liu, C. Xu, H. Guo, and Y. Liu, "A survey on emerging computing paradigms for Big Data," Chinese J. Electron., vol. 26, no. 1, pp. 1–12, 2017, doi: 10.1049/cje.2016.11.016.

[24] I. A. T. Hashem, I. Yaqoob, N. B. Anuar, S. Mokhtar, A. Gani, and S. Ullah Khan, "The rise of 'Big Data' on cloud computing: Review and open research issues," Inf. Syst., vol. 47, pp. 98–115, 2015, doi: 10.1016/j.is.2014.07.006.

[25] J. Xu, E. Huang, C. H. Chen, and L. H. Lee, "Simulation optimization: A review and exploration in the new era of cloud computing and Big Data," Asia-Pacific J. Oper. Res., vol. 32, no. 3, pp. 1–34, 2015, doi: 10.1142/S0217595915500190.

[26] A. Oussous, F. Z. Benjelloun, A. Ait Lahcen, and S. Belfkih, "Big Data technologies: A survey," J. King Saud Univ. - Comput. Inf. Sci., vol. 30, no. 4, pp. 431–448, 2018, doi: 10.1016/j.jksuci.2017.06.001.

[27] P. Pedamkar, "Cloud computing vs Big Data Analytics." https://www. educba.com/cloud-computing-vs-big-data-analytics/ (accessed Sep. 24, 2022).

[28] S. Shrivastava, "Big Data and Cloud computing: A Winning Combination! - Ksolves Blog." https://www.ksolves.com/blog/big-data/ big-data-and-cloud-computing-a-winning-combination (accessed Sep. 24, 2022).

[29] Krishnaveni, N., Nagaprasad, S., Khari, M., & Devi, V. A. (2018). Improved Data integrity and Storage security in Cloud Computing. *International Journal of Pure and Applied Mathematics*, *119*(15), 2889–2897.

[30] M. Bahrami and M. Singhal, "The Role of Cloud computing Architecture in Big Data," Stud. Big Data, vol. 8, pp. 275–295, 2015, doi: 10.1007/978-3-319-08254-7_13.

[31] C. Yang, Q. Huang, Z. Li, K. Liu, and F. Hu, "Big Data and cloud computing: innovation opportunities and challenges," Int. J. Digit. Earth, vol. 10, no. 1, pp. 13–53, 2017, doi: 10.1080/17538947.2016.1239771.

[32] A. Malhotra, "Big Data and Cloud computing – A Perfect Combination." https://www.whizlabs.com/blog/big-data-and-cloud-computing/ (accessed Sep. 24, 2022).

[33] R. Delgado, "The pros and cons of harnessing Big Data in the cloud | ITProPortal." https://www.itproportal.com/2013/12/20/pros-and-cons-of-big-data-in-the-cloud/ (accessed Sep. 24, 2022).

SECTION 2

Advance Implementation of Artificial Intelligence and Data Analytics

7

Artificial Intelligence-based Data Wrangling Issues and Data Analytics Process for Various Domains

V. Pandimurugan[1], V. Rajaram[1], S. Srividhya[2], G. Saranya[1], and Paul Rodrigues[3]

[1]Department of networking and communications, School of computing, SRMIST, kattankulathur, Chennai
[2]Department of DSBS, School of computing, SRMIST, kattankulathur, Chennai
[3]Professor, Computer Engineering, Collge of engineering, King Khalid university, Abha, Saudi Arabia.

Email: vpcoe84@gmail.com; rajaramv@srmist.edu.in; srividhs1@srmist.edu.in; saranyag3@srmist.edu.in; prigues@kku.edu.sa

Abstract

In day-to-day life, data plays a vital role, and it transforms from information to knowledge, wisdom also provides an opportunity in digital technology. All firms that want to accomplish this ambidexterity should depend on data as a fundamental resource for their operations and create data-driven business models. Applying several functions to the data, such as collecting, analyzing, and displaying it while utilizing all levels of knowledge can help build new products and customers, and reach goals at the appropriate moment. Due to the large volume of data increasing drastically, handling data is crucial for humans. Artificial intelligence allows machines to mimic human brain functions and is also used to identify data types, find relationship between datasets, and recognize knowledge. It also performs explore data, speed up data preparation, automation, and creation of data models. AI in data analytics can uncover ideas, discover new patterns and establish new relationship among the data. NLP plays an important role in data analytics for extracting data, text classification based on the keyword, sentiment analysis, and also AI predictions are accurate, because of access to huge amounts of data, enough power of data, and ML detects patterns and learns. Data wrangling is the process of handling complex data, involving various operations such as discovering, structuring, cleaning, validating, and also enriching the data for betterment. AI is giving the best solution for data wrangling and analytics in various domains, such as marketing, business, healthcare, banking, finance, etc.

7.1 Introduction

Accurate, timely, specific, and organized data can increase understanding and decrease uncertainty, and can be presented within a context that makes it meaningful and relevant. Information refers to data that is specialized and organized with a goal, as opposed to data, which are unprocessed facts and statistics. The term information also refers to data that has been evaluated before being put in a more meaningful context. It gives a firm a basis for decision-making.

7.1.1 Need for information inside the organization

For a firm to make wise decisions, information must be presented in a way that management can understand [1]. It would be helpful to have customer data in this situation it can be used to develop better ways of engaging or collaborating with our clients. Information, however, must also be acknowledged for its value in inspiring behavior. A company's response to consumer information is only useful if it prompts a corresponding change in how it interacts with consumers. If you receive a warning about low customer satisfaction, what should you do? The information process should be a part of a bigger internal review process to produce the best results.

In many occupations, data is crucial for understanding activity metrics, professional performance metrics, economic performance metrics, and operational performance measurements. On a theoretical level, data is defined as unprocessed outputs, whereas information is defined as data that has been transformed into insight. The greatest way to understand the career and how it operates is to use this knowledge. The ability to better satisfy the client's needs by learning more about them is one of the most important applications of information in the workplace. The client is at the center of everything that one can do, regardless of profession or line of work. To communicate with the customer, one must be able to comprehend them, to encourage them to interact and one can interact with the customer on equal footing by embracing data.

7.2 Categories of Data

Data is the future of oil. In today's world, data are everywhere. Data scientists, marketers, businesspeople, data analysts, researchers, or anyone else who works with raw or structured data must play with it or experiment with it regardless of their profession. Because the information is so crucial, it must be handled and stored correctly and error-free. To analyze these data and provide the desired outcomes, it is crucial to understand the different forms of data [2]. The two types of data that can be categorized are qualitative and quantitative, which are further separated into nominal, ordinal, discrete, and continuous categories.

Nowadays, businesses rely heavily on data, and the majority of them utilize data-driven insights to establish strategies, introduce new goods and services, and test out new ideas. Today, at least 2.7 quintillion bytes of data are generated each day, according to research qualitative or categorical data cannot be quantified or tallied numerically. It is categorical rather than quantitative data that is arranged in this way. It also goes by the name categorical data for this reason. These data can be text, symbols, audio, or pictures [3].

7.2.1 Qualitative data

Qualitative data describes how individuals see things. Market researchers may use this information to better understand the preferences of their target market and then adapt their ideas and approaches [4].

Two additional categories have been added to the qualitative data:

Nominal data
    Variables with no order or numerical value are labeled using nominal data. Since one hair color cannot be compared to another, hair color might be thought of as nominal data.

    Illustrations of nominal data

    • Color of a bike (white, half-white, black, and red)

    • Nationality (Indian, German, and American)

Ordinal data
    Ordinal data have a natural ordering in which the numbers are arranged in some way according to their scale positions. These statistics are used to track things like consumer pleasure and satisfaction, but we cannot do any mathematical operations on them.

An ordinal format for qualitative data has values that are positioned in relation to one another. These data can be considered to be "in-between" qualitative and quantitative data. Ordinal data only shows sequences; hence it cannot be utilized for statistical analysis. In contrast to nominal data, which lack any type of order, ordinal data show some sort of order.

7.2.2 Quantitative information

Quantitative data can be described by numerical values, which allows for counting and statistical data analysis. This information is also known as numerical data. Inquiries like "how much," "how many," and "how often" are addressed. Quantitative information includes items like a person's height or weight, the price of a phone, the RAM in a computer, etc.

A wide variety of graphs and charts, including bar graphs, histograms, scatter plots, boxplots, pie charts, and line graphs, among others, are used to visualize and manipulate quantitative data.

Quantitative data examples
- The price of a bike

- The boiling point of liquid

Categories of the quantitative data

Separate data

Discrete refers to anything unique or separate. The values that fall under integers or whole numbers are contained in the discrete data. Discrete data include things like the overall number of pupils in the class. There is no way to convert this data into decimal or fractional numbers.

Since the discrete data are countable and have finite values, they cannot be divided. These data are typically represented using a bar graph, number line, or frequency table.

Examples of discrete data include the number of employees at a corporation.

- The price of a watch

- Number of students in a class

Continuous data

Fractional numbers are the representation of continuous data. It may be an Android phone's version, someone's height, the size of an object, etc. Information that may be broken down into lesser levels is represented by continuous data. Any value within a range can be assigned to the continuous variable.

The main distinction between discrete and continuous data is the presence of the integer or whole number in discrete data. The fractional values are still stored in continuous data to record many forms of data, including temperature, height, width, time, speed, etc.

A few instances of continuous data

- Dimensions of a person

- Speed of sound

7.3 Why AI in Data Analysis

AI analytics automates the time-consuming work that a data analyst would traditionally do by using machine learning programs to continuously monitor and analyze enormous volumes of data [5].

Business intelligence and analytics

Following the scientific advancements of AI over the past several years, it has had a significant influence on practically every business. By enabling levels of velocity, size, and granularity that are difficult for humans to reach, AI and machine learning are revolutionizing the analytics field.

Analytics refers to the identification, interpretation, and communication of relevant data trends. Business intelligence relates to the use of this methodology to analyze data, estimate future outcomes, find new connections, and ultimately improve decision-making [6].

Analytics, in its simplest form, is the process of applying an analytical framework to unstructured data to spot important trends. There are many different analytical techniques, but the following are some of the most used:

- Mathematical applications

- Statistical analysis

- Artificial intelligence and machine learning

According to the gartner analytic ascendancy model, the area of analytics may be further subdivided into many levels.

Image origin

There are four phases to the gartner logical ascendancy model, each with a higher level of complexity and significance (Figure 7.1).

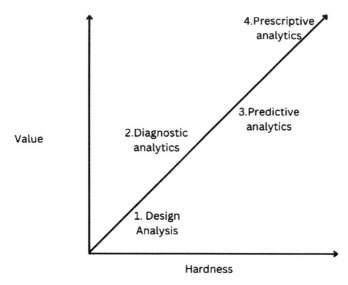

Figure 7.1   Phases of complexity and value.

- Descriptive analytics: The analyst must determine what has already happened in the data during the initial phase of analytics, which is relied on hindsight. Diagnostic analytics: The next stage is insight-driven and calls for the analyst to ascertain the reason behind a specific incident or data modification.

- Diagnostic analytics: Oftentimes, the most challenging and valuable step in analytics is identifying precisely how to achieve the intended objective.

Predictive analytics: The next stage in analytics is focused on foresight and predicting what will occur in the future.

Artificial intelligence (AI)

Computer science's broad field of AI includes all varieties of artificial intelligence. Computers that replicate cognitive processes like training, problem-solving, logic, and visualization are frequently referred to in this context. Anything can be done with AI, including voice recognition, autonomous driving, video games, and, of course, analytics. Artificial neural networks, statistical techniques, and search engine optimization are a few examples of AI-based problem-solving techniques.

Automatic learning
Machine learning is a branch of AI technology that performs a task using methods, mathematical analysis, and facts without being explicitly programmed.

Machine learning models rely on patterns and interpretation rather than providing detailed instructions on how to finish a task. Machine learning specifically calls for the development of a model that is taught using training data and can then be given fresh data to produce predictions.

Analytics AI
AI analytics, a subfield of business intelligence, uses machine learning techniques to discover unique patterns, relationships, and concepts in data. In reality, AI analytics is the process of automating a large portion of a data analyst's duties. AI analytics improves a data analyst's capabilities in terms of quickness, the amount of data that can be analyzed, and the level of tracking delicacy. While analysts are not intended to be replaced, AI analytics typically boosts the volume and speed of data that analysts can analyze.

AI analytics vs conventional analytics
Traditionally, a technical team of data analysts is responsible for data analytics. Here is an example of how historically a team of analysts can approach a business challenge: Over a period of time, an event, occurrence, or trend occurs in the institution – for example, more students fail a specific course. Data analysts then formulate hypotheses regarding the likely causes of the sales decline. Then, these hypotheses are compared to the facts for that time period until sufficient evidence is found to support a certain theory [7].

It is understood that the entire procedure, from identifying the initial alteration to identifying its root causes, is incredibly time-consuming. In addition, there is no assurance that the answers found by data analysts are correct owing to their inherent constraints.

On the contrary side, machine learning-based AI analytics continuously track and analyze huge amounts of data. Following are some ways that the results are different from traditional analytics [8]:

- Scale
  An AI-based solution for anomaly detection learns the usual behavior of the data without being specifically instructed on what to look for. It does it at any level of detail, including revenue by nation, product, and channel.

- Speed
  The AI model will detect abnormal revenue declines and immediately notify the necessary teams. Additionally, a clustering and correlation technique is used by an AI-based analytics system to provide a root cause study, enabling quick resolution of any issues. Because the analysis is done constantly and in real time rather than quarterly, monthly, or

at most weekly as is the case with conventional analytics, the remedial time is reduced by orders of magnitude.

- Accuracy
  A human analyst would find it almost impossible to identify and provide relationships between anomalies using the ML algorithms on which AI analytics is based (correlations between millions of time series in some cases). Naturally, the design of the ML algorithms determines their quality; they must correctly acquire a range of patterns automatically, which requires the use of numerous techniques.

  Moreover, unlike data analysts, these algorithms have no preconceived notions about the business problems at hand. For instance, AI analytics can evaluate enormous volumes of data and give a totally objective appraisal of the issue, as opposed to relying on preexisting assumptions about the potential causes of a shift in income.

- E-commerce reports
  This is a fantastic use case for AI analytics in e-commerce because there are so many data points that could be influencing changes in conversion rates. Additionally, having a system that continuously analyses data enables you to identify issues early on because this is a problem that e-commerce organizations must constantly battle. It can prevent the company from losing a significant sum of money. To be more precise, an AI-powered system could learn the subtleties of your conversion rate, spot changes on its own, and produce real-time projections.

- Financial technology analytics
  The detection and prevention of possible security concerns are examples of prescriptive analytics in FinTech. By monitoring the behavior of operational metrics, AI analytics may be used to address security gaps so that it can be proactive about security. By consolidating various data sources into a single platform, machine learning may be used to comprehend how these metrics operate typically, discover abnormalities, and avert problems in real time.

- Telecom intelligence
  In the telecommunications business, one example of AI analytics is answering queries such as "is the wifi network stable?"

The above issue may be answered by artificial intelligence by automatically recognizing changes in service quality, which can also minimize churn and boost Average revenue per unit. In particular, an AI solution such as autonomous business monitoring and anomaly detection may leverage its root cause analysis and correlation engine to minimize the amount of time required to resolve possible network issues.

At the start: AI Offers a Competitive Advantage to the Contemporary Analytics Stack Creating hypotheses, preprocessing data, visualizing data, and employing statistical techniques are examples of tasks that traditionally require a significant amount of human effort.

The problem with this strategy, however, is that the time required to conduct these procedures manually is just too long for the fast-paced commercial environment of the present day. For this reason, many businesses are using AI analytics for demand forecasting, anomaly identification, and company monitoring. Particularly, machine learning algorithms may be utilized to boost the skills of your technical staff so that they can adapt immediately to business developments.

7.4 Why AI Is Required for Data Manipulation?

7.4.1 Data wrangling

Data wrangling is the process of collecting, classifying, and processing "raw" data in preparation for analysis and other downstream operations. It is distinct from data cleansing since it goes beyond only deleting erroneous and unnecessary data and transforms, reformats, and prepares it for potential downstream needs in a more comprehensive manner [9].

"Wrangling" because it involves managing and manipulating data lassoing it, if you will so that it is better gathered and structured for whatever purpose it will serve. It converts the data from its original format into the format(s) required for the need.

Despite significant advances in AI and machine learning, data manipulation remains mostly manual, or humans must at least coach computer programs through the procedure. AI is not yet sufficiently developed to be "let loose" on raw data to modify it for downstream purposes. However, a variety of open source and Software as a service have greatly simplified the procedure.

Data manipulation demands a sophisticated understanding of the raw data, the analysis that will be performed on it, and the information that must be eliminated. A variety of precise regulations and guidelines must be implemented to guarantee that the data is cleaned and prepared appropriately; otherwise, your analysis will be biased or altogether ineffective (Figure 7.2).

Figure 7.2   Various tasks in data wrangling.

- Eliminating useless data from the analysis. In text analysis, they may include stop words (such as the, and, etc.), URLs, symbols, and emojis, among others.

- Eliminating data gaps, such as empty cells in a spreadsheet or commas between sentences in a text.

- Integrating data from numerous sources and sets into a single data set.

- Filtering data based on regions, demographics, time periods, etc.

These are the operations that will prepare the outputs for subsequent demands. These can be automated operations, but the data scientist or human executing the wrangling must know which types of activities are pertinent to their needs and order the algorithms to carry out those specific tasks.

7.4.2 Why is data wrangling important?

If the analysis is not done using clean, accurate data, your conclusions will be biased, frequently to the point of being more damaging than useful. Essentially, the more accurate the input data, the more accurate the output results will be. In addition, as the amount of raw data you use increases, so will the amount of "bad" or unneeded data contained inside these data sets. However, if you have a solid data wrangling procedure, you will be able to clean this data more quickly as you advance through it because you will have processes in place to manage it.

Data scientists will frequently assert that having better data is more essential than having the most powerful algorithms, and that data manipulation is the most crucial phase in data analysis. If you do not begin with quality data, your subsequent procedures are essentially meaningless [10].

7.4.3 Data cleaning procedure

There are six phases to data manipulation that are universally acknowledged by the data science community:

- Exploring
  Discovering is the initial stage in data manipulation; it involves gaining an overview of the data. What do your data consist of? Consider how you may utilize your data and what insights you might glean from it as you become acquainted with it.

  When you examine or read through it, you will begin to see trends and patterns that will help you comprehend the type of study you may like to do. You will also likely identify immediate faults or bad data items that need to be corrected or eliminated.

- Structuring
  Next, you must arrange or organize your data. Some data, such as a hotel database's database, is completely inputted into spreadsheets and arrives prestructured. However, some data, such as open-ended survey replies and customer assistance data are unstructured and unusable in their raw form.

  The act of preparing your data so that it is homogeneous and suitable for analysis is known as data structuring. How you arrange your data will rely on the original format of your data and the analysis you intend to do – the downstream procedures.

- Sanitation
  The act of eliminating extraneous data, mistakes, and inconsistencies that might bias your results is known as data cleaning. The process of data cleansing involves deleting empty cells or rows, removing blank spaces, repairing misspelled words, and reformatting dates, addresses, and phone numbers so that they are all uniform. Errors that might adversely affect your subsequent analysis will be eliminated using data cleansing technologies.

7.4.4 Enrichment of data fine-tuning

After gaining an understanding of the data you are dealing with and correctly converting and cleaning it for analysis, it is important to determine whether or not you have all the data required for the current task. Occasionally, following the data cleansing process, you may discover that you have far less useful data than you originally believed. Data enrichment integrates an organization's internal data from customer service, CRM systems, sales, etc., with data from third parties, external sources, social media input, and pertinent web-wide data.

In machine learning, the basic rule is that more data is better. However, this will depend on the current project. If you desire to add additional data, simply run it through the aforementioned steps and append it to your existing data. Enriched data will facilitate the development of individualized interactions with clients and give a more comprehensive knowledge of their demands. When a corporation fails to enrich its data regularly, it will have a worse grasp of its consumers and supply them with irrelevant information, marketing materials, and products.

- Validating
  Data validation is the process of verifying and proving the standardization, consistency, and quality of your data. Verify that it is neat and well-organized. Checking the quality and validity of cells or fields by cross-checking data often involves the use of automated tools, although it may also be performed manually.

- Publication
  Now that your data has been confirmed, you may publish it. This might involve sharing information throughout your enterprise or organization for various analytic purposes, or uploading it to machine learning algorithms to train new models or run it through already learned models. Depending on your needs and objectives, you may distribute data in a number of formats, including electronic files, CSV documents, and written reports.

These six processes are a continuous procedure that must be performed on every new data to maintain its cleanliness and readiness for analysis. Let's have a look at various approaches and tools for manipulating data to achieve the best outcomes.

7.4.5 Data wrangling methods and instruments

One of the finest qualities of a competent data wrangler is intimate familiarity with their data. You must comprehend the contents of your data sets, your data analysis tools, and your final objectives. Historically, the majority of data manipulation was performed manually using spreadsheets such as Excel and Google Sheets. Typically, this is a viable option for tiny data sets that do not require extensive cleansing and enrichment.

When dealing with higher volumes of data, it is usually best to utilize more complex technologies. Several code toolboxes may be incredibly useful for data-wrangling chores if you know how to code. Scalability is facilitated by wrangling tools such as KNIME, but you must be fluent in languages such as Python, SQL, R, PHP, and Scala.

There are excellent SaaS data manipulation solutions that do not require coding. Even developers are gravitating to no-code or low-code solutions due to their user-friendliness and intuitive interfaces. Those who wish to execute some code or more in-depth data manipulation are still able to write their own code within these tools [11].

7.4.5.1 Factual wrangler

Trifacta Wrangler is a fully integrated data wrangling tool that evolved from Stanford's initial data wrangler. Trifacta's user-friendly dashboard allows both noncode users and coders to collaborate on enterprise-wide projects. And its cloud scalability enables Trifacta to be applied to projects of any size and to automatically export manipulated data to Excel, R, Tableau, and more.

7.4.5.2 Open refine

Open Refine, formerly a SaaS-only tool formerly known as Google Refine, has extended a hand to the coding community to make improvements as they see appropriate. Its effective and intuitive graphical user interface enables users to view, investigate, and clean data without writing any code. However, Python's features allow you to conduct even more intricate data filtering and cleansing using your own code.

7.4.5.3 Bottom line

Final Verdict, Data wrangling is the new frontier and the new industry standard in the collection, cleaning, and preparation of data for analysis. When properly implemented, it could help ensure that you use the best, most understandable data available and position you for productive downstream operations.

Wrangling or data munging can be a tiresome procedure, but you will be glad you added it to your arsenal of data analytics tools once you see the results.

7.4.5.4 MonkeyLearn

It is a SaaS text analysis platform with robust capabilities for extracting maximum value from text data. Examine MonkeyLearn to discover how we may assist you in optimizing your data.

AI in data wrangling

Transforming between formats

Data transformation and reduction are fundamental to artificial intelligence (AI) and machine learning. Francis Galton, an Englishman, measured the circumference of 700 sweet peas 140 years ago to determine whether or not pea size is inherited. He developed linear regression along the way, a technique to machine learning that is still widely used today (unlike conventional least squares or R2, which had existed for some time). Galton was more concerned with one aspect: whether something was hereditary or not, rather than 700 circumferences.

Database marketing gained popularity a century later, and today, programs like Experian CustomerView and Merkle M1 allow you to access 10,000 pieces of information about every adult in the United States. Analysts, however, desire one number per person rather than 10,000 data points, such as a person's propensity to churn or get pet insurance. On HBO's Silicon Valley, deep learning was recently portrayed as reducing digital photos to a single bit to millions of viewers.

Tens of thousands of Pdf documents detailing their current and prior products, as well as those of their rivals, were in the possession of a global manufacturer of food additives. People benefit from PDFs, but XGBoost or TensorFlow does not. It is necessary to determine each product's chemical makeup, melting point, viscosity, and solid fat content at room temperature. They could have done it manually by hiring a large number of foreign readers, but that would have been slow, inaccurate, and expensive, all things that data scientists abhor. Instead, they created an algorithm to quickly and automatically extract the data from every PDF.

The second example also has audio files. Digital twins for factories are a hot topic right now, but because there are frequently many stakeholders involved, many of whom lack IT infrastructure expertise, adding digital sensors to large, outdated facilities can be slow, inaccurate, and expensive (as mentioned above, three things data scientists abhor). Some people take the shortcut of installing hundreds of wireless temperature, sound, and vibration sensors throughout a site to create a digital twin that is at least minimally functional. For example, a temperature sensor on a pump can show if it is operating unusually hard, which may point to a clogged pipe.

In a pump, squeals that indicate bearing failure may be heard by sound sensors. It is possible to stream sensor data to cloud folders for processing. The plant maintenance manager, however, does not need 20,000 audio files. They merely need to know which pumps are most likely to fail and when. The task of turning thousands of audio samples into a few hundred probabilities and dates is ideally suited for AI.

Human inspectors used to be in charge of this, but they are slow, inaccurate, and expensive. Their accuracy varies greatly depending on the hour of the day, the day of the week, and the month. They developed AI technologies that condensed each photo to a single bit, flaw or not and provided real-time access to human inspectors. In one instance, inspectors and AI worked together to reduce false positives by 50% and find 98% of the mistakes.

7.5 Data Science Lifecycle

The core of the data science lifecycle is the application of machine learning and numerous analytical approaches to develop insights and predictions from data to achieve a business goal. The overall method is made up of several steps, including data preparation, cleaning, modeling, and model evaluation. It is a lengthy procedure that could take several months to complete. Therefore, having a broad framework to work within for any issue at hand is crucial. The CRISP-DM framework, also known as the Cross Industry Standard Process for Data Mining, is the globally accepted framework for resolving any analytical problem [12].

Previously, data was considerably more limited and generally available in a well-structured format, which we could save quickly and readily in Excel sheets. With the aid of business intelligence tools, data can be rapidly analyzed. Every day, around 3.0 quintillion bytes of records are created, resulting in a data explosion. Recent studies suggest that a single human generates 1.9 MB of data and recordings every second.

Managing the massive amount of data that is generated every second is a tough issue for any corporation. To handle and analyze this data, we required very sophisticated tools and algorithms, which is where data science comes in. The following is a summary of some of the most important justifications for employing data science technologies:

- It assists in turning a sizable amount of unstructured, raw data into insightful knowledge.

- It may support specialized projections, including a range of polls, elections, etc.

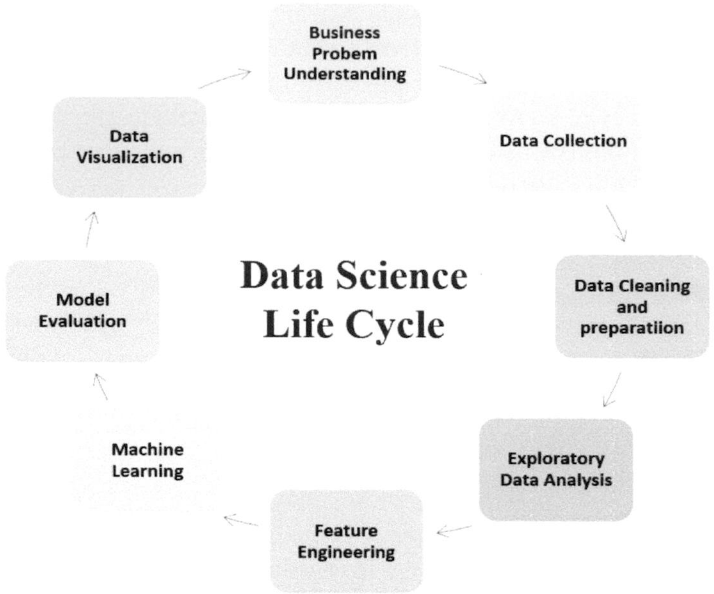

Figure 7.3 Data science life cycle.

- It contributes to the automation of transportation, including the creation of self-driving cars, which are the wave of the future. Data science is being embraced by businesses that favor this technology (Figure 7.3). Data science techniques are used by companies like Amazon, Netflix, and others who manage large amounts of data to enhance the user experience [13].

7.5.1 The evolution of data science

7.5.1.1 Business acumen
The corporate objective is the center of the entire cycle. What will you fix if you don't have a specific problem anymore? The business purpose will serve as the study's main goal, thus it is crucial that you fully understand it. We can only choose a specific evaluation target that is consistent with the business goal after gaining a favorable perspective. You must ascertain the client's preference for either reducing savings loss or estimating a commodity's rate, among other things [13].

7.5.1.2 Data understanding
Data comprehension is the next level beyond enterprise comprehension. This includes a list of all the data that is available. Since they are knowledgeable

about the available information, the data that should be used for this corporate challenge, and other information, you must work closely with the enterprise group in this situation. The structure, importance, and record type of the data must be specified at this stage. Analyze the data via graphical displays. Basically, finding any informational facts about the data that can be discovered through straightforward data exploration.

7.5.1.3 Data preparation

The preparation of the data comes next. This involves choosing the appropriate data, integrating it by combining data sets, cleaning the data, dealing with missing values by either removing them or imputing them, handling inaccurate data by removing them, and testing for outliers using box plots and handling them. By generating new elements from existing ones, you can create new data. Remove any unnecessary columns or features before formatting the data as desired [18]. The most time-consuming but possibly most important phase of the existence cycle is data preparation. As accurate as your data is, so is your model.

7.5.1.4 Exploratory data analysis

Prior to creating the actual model, this method entails getting a general grasp of the solution and its impacting aspects. The distribution of data inside a character's numerous variables is examined using bar graphs. Heat maps and scatter plots are used to show how several factors relate to one another. To recognize each attribute both on its own and when combined with other qualities, a variety of data visualization approaches are widely used.

7.5.1.5 Data modeling

Data analysis depends on it fundamentally. A model uses the organized data as input and produces the intended output. This stage involves choosing the proper kind of model, regardless of whether the task is a classification problem, a regression problem, or a clustering challenge. From among the many algorithms contained inside the model family, we must carefully choose which ones to enforce and apply. To get the desired performance, the hyperparameters for each model must be optimized. We also need to make sure that generalizability and overall performance are balanced. The model should no longer analyze the data and perform poorly with new data [14].

The model is assessed in this case to determine whether it is prepared for deployment. Unobserved data are used to assess the model, and it is then assessed using a set of assessment measures that have been carefully created. Furthermore, we must make sure that the model is accurate. The entire modeling technique must be repeated until the necessary level of metrics is reached if the evaluation does not produce a suitable result. Any data science solution or machine learning model must evolve, be able to get better with new data, and adapt to a new assessment metric, just like a human [19, 20]. For a given phenomenon, several models might be developed, but many of them might also have errors. The examination of the models makes it easier to choose and build the best model.

7.5.1.6 Model deployment

The model is finally implemented in the desired structure and channel following a thorough evaluation. completing the data science life cycle. The aforementioned data science life cycle must be meticulously carried out during each stage. Any step that is carried out improperly will have an impact on the following stage and render the entire effort useless. For instance, if data collection is no longer efficient, records will be lost, and the creation of an ideal model will cease. The model will not work properly if the data is not correctly cleansed. The model will fall short in the real world if it is not properly evaluated. It takes enough time, effort, and attention to complete each stage, from the conception of the business to the implementation of the model.

7.5.2 Enhancements to data analytics

As technology advances swiftly, artificial intelligence (AI) continues to excite and captivate individuals. It contributes to analytic skills by making data more accessible, automated, and analyzable. AI describes a machine's learning, logic, reasoning, and perception, which were once thought to be human-specific but are today copied by technology and employed in virtually every industry. It employs computer science programming to stimulate human cognition and activity by assessing data and surroundings, resolving or predicting difficulties, and adapting to a range of jobs. Applications of artificial intelligence include big data, face recognition, cybernetics, robotics, drones, 5G, smartphones, cryptocurrency, autonomous weaponry, and information sharing [15].

According to the Brookings Institute, AI technology is revolutionizing all aspects of society and has several uses, such as data processing, decision-making, and the transmission of knowledge.

Several contributions of AI and machine learning to analytics:

- AI automates report development and simplifies data comprehension.

- Artificial intelligence streamlines procedures, accelerating the generation of insights.

- Using machine learning algorithms, AI examines data to forecast future events and show trends and patterns.

- Artificial intelligence reduces mistakes and provides more accuracy than conventional business intelligence solutions.

Industrial impact of artificial intelligence

How is AI altering the global landscape? AI is altering the lives of ordinary people and several sectors. The influence of AI may be detected everywhere. Health care, finance, logistics, and travel are just a few of the industries that employ AI to deliver an exceptional experience.

Some examples of how big businesses use AI include:

- Transportation. In several places of the world, autonomous cars are already in use. In the meanwhile, AI is utilizing data in traffic management to increase road safety and in transport services, including air travel, to decrease delays and wait times.

- Manufacturing
Several industrial procedures are aided by machine learning models. Because manufacturing is replete with analytic data, AI technologies can foresee the need for equipment maintenance, manage stocks, and design and enhance goods.

- Medical care
The use of data analytics has had a significant influence on health care. Quality indicators from a multitude of interconnected digital health instruments have enhanced the patient experience [21, 22]. Personalized care can be based on an individual's genetic information.

- Banking
Chatbots strengthen customer connections and provide personalized assistance, insights, and suggestions. AI avoids fraud by recognizing out-of-the-ordinary expenditures and highlighting those that do not conform to typical patterns or criteria.

- Travel. AI has enabled seamless travel for individuals. Machine learning and data science enable the adaptation of dynamic and real-time content to locales and consumer behaviors, presenting the traveler with the most pertinent information.

7.6 IBM Watson's Role in Data Analytics

Watson Analytics indicates that Time of Day and Weekday have a 66% ability to forecast call volume. Indeed, it is accurate. In my dataset, call volumes are highly dependent on:

- Time of Day – there are many calls during morning and evening office hours, and fewer calls during sleeping hours;

- Weekday – there are many calls on weekdays and very few calls on weekends; and

- Day of the Week – there are many calls on weekdays and very few calls on weekends.

With IBM Watson analytics, you can ask business questions and get answers with a single click using predictive analytics and natural language processing. With a single line of code, you can import Twitter data in real time and analyze public mood as an event unfolds [16].

7.6.1 How IBM analytics operates

Databases can be directly linked or uploaded as Excel or CSV files to Watson analytics. As an alternative, before beginning your study, you can import Twitter data directly into WA's cloud storage. You can start asking direct questions to WA once your data has been uploaded to the cloud, and WA will respond with data visualizations that best address your inquiries. You can dig deeper into these graphs and analyze the data by posing new queries or coming up with your own hypothesis. These analyses are what Watson Analytics refers to as "discoveries."

Communication of results is a crucial component of knowledge discovery, and WA gives several methods for doing so, such as sending an email, tweeting the results, downloading an image/PDF/PowerPoint, or sending a link to a colleague. Watson Analytics is supported by several well-established principles, including [17].

- Natural language processing

- Cloud computing

- Predictive analytics

- Data visualization

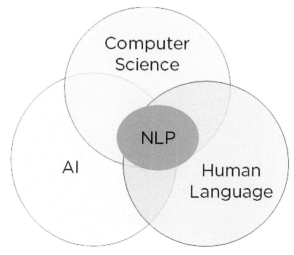

Figure 7.4   Constituents of NLP.

7.7 Role of NLP in Data Analytics

NLP is a crucial component of AI that allows it to do any task effectively. NLP employs artificial intelligence to take actual input, interpret it, and make sense of it in a way a computer can comprehend. The machine will understand and process natural human language here, whether the language is spoken or written. NLP is used in many different ways in AI, including speech recognition, machine translation, question-and-answer functions, and personal help [23]. In NLP voice is very important, because as humans, text is meaningful information, but for machines, it is about strings and symbols, so we have to train various levels and algorithms to teach the machine to understand the meaning.

The world is now connected globally due to the advancement of technology and devices. Due to this advanced technology tons of data are generated, it is difficult to analyze the various forms of data like text, image, audio, and video and also handle different languages (6500 languages globally). NLP combines the field of computer science, AI and human language (Figures 7.4 and 7.5).

NLP performs based on different steps like segmentation, tokenization, stemming, and lemmatization. First, the text should be involved in the segmentation process, which needs to break the entire document into its constituent sentences. We can use special characters like punctuation, commas, etc. in the constituent sentences. For the tokenization process, the sentence is split into words, each word may have been called a token. Stem is the important

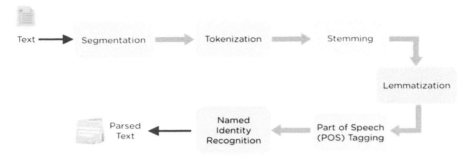

Figure 7.5   NLP performance steps.

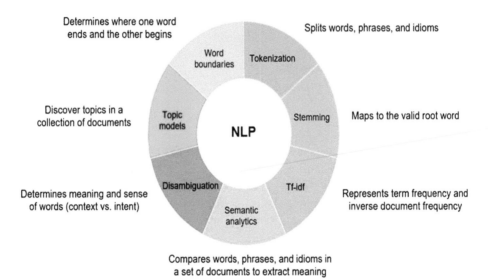

Figure 7.6   NLP performance.

terminology in the stemming process to obtain the word [28]. If new word wants to create, we can add the affixes to the existing and create the new word. In the dictionary, all the information will be present, how the word arrives and derived from which origin and the other factors like tense, mood, gender, etc. (Figures 7.6 and 7.7).

Parsing is the vital task for NLP to complete the process, whether it may be a structure of string or text. The sentence has many structures, to form a new sentence, word, or phrase, but it should be meaningful for the specified application. Building the sentence is not a simple task, but by using the NLP technique we can make it simple and meaningful. Processing NLP has many challenges, the final task is to train the model for converting a piece of text

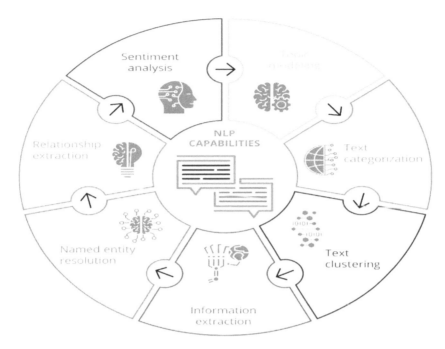

Figure 7.7 NLP capabilities.

into a machine-understandable one [24]. The NLP tasks include the following, in the order of going from raw text to classified text for understanding:

Part of speech tagging (POS): In this process, the syntax, such as nouns, adverbs, and verbs, among others, must be identified.

(CHUNK) Chunking It is the process of locating word chunks like verb and noun phrases.

Name entity recognition (NER) identifies the entity associated with the tag, such as a person, business, or location.

SRL: Semantic role labeling The hardest part of this endeavor is determining the semantic functions of the words and tags for the right phrase to tag.

7.7.1 Tools for data analytics and visualization

Visualization of data is the process, in which a huge amount of data is presented in a graphical format with easily understandable tags. Data is in the raw format is not used for many applications because it needs some preprocessing techniques to streamline the data for practical and research use. Data collected from the various applications, basics refinement steps to be followed like data cleaning, data remembrance, etc. [25]. Analytics of the data is the art to classify which analysis is suitable for what type of data and which is the most suitable application. Data analytics can be classified as predictive, prescriptive, diagnostic, and analysis. Data analytics is important for business process models because then only we may know what the real problem is, how we can design the prototype model, etc.

Nowadays, in many IoT applications, data generated are involved in any of the above said analytics methods to refine the application in a better way to reach the customer.

For descriptive analytics, it should be analyzed over a period of time that these types of analyses may be applied mostly in the business intelligence model.. Diagnostics analytics are focused on why its happening, and the reason to find the problem or issue. Mostly these types of analytics are used in the healthcare sectors [27]. Predictive analytics are focused on the future, and what may happen, if a problem arises in the model, and mostly it will be applied in the training design model applications.

7.7.1.1 Tools for data analytics

Splunk is used for analyzing the machine-generated data and websites efficiently. It has products like Splunk Free, enterprise, and cloud for the various applications. Talend is the most useful data analytics tool that aims to deliver clean data to everyone. Data is used with the Qlikview tool, which offers features like data integration and data literacy. The Power BI tool is used so that end users can build their own dashboards and reports independently [26]. Working with a live data set and spending more time on data analysis rather than data wrangling are both made possible by Tableau.

7.7.1.2 Uses of data visualization

Data visualizing is the process of a huge amount of data or information, whether it may be text, or any format, can convert and show a graphical format, which can be easily understood by all end users. It is also defined as the collected, processed data, and model to be finalized, when it is come to a conclusion, and at that time we can make data visualizing. Data visualization can be included in the ability to absorb the information quickly and understand, easy distribution of the resource, and information for more visibility [29].

General types of visualizations:

- **Chart**: By using two axes can draw the information in a graphical manner such as bar, pie, histogram, dotted graph, etc.

- **Table**: Information available in the format of rows and columns.

- **Graph**: By using variables or points, we can represent the information in the format of lines, segments curves, etc.

- **Geospatial**: It shows the data based on the locations and shapes, and for classifying using colors [30].

- **Infographic**: It combines visuals and data.

- **Dashboards**: A group of data visualizations and data displayed in one location to aid in data analysis and presentation.

- Additional examples

- **Area map**: This displays data visualization by state, country, or any other geolocation.

- A bar chart should be drawn based on the numerical values. Each variable's value is represented by the length of the bar.

- **Box-and-whisker plots**: Displays the graph over the specified measure. It could be a box, a bar, or a selection range.

- **Bullet graph**: A bar marked against a background to demonstrate performance in relation to a target, indicated by a line on the graph.

- **Gantt chart**: This timeline-based chart is most frequently used in project management software.

- **Heat map**: A type of geospatial visualization in map form that uses different colors to represent different data values (temperatures are not always the case, but that is a common use).

- **Highlight table**: Its data is presented in different categories with contrasting colors.

- **Histogram**: A type of bar graph that divides a continuous measure into various bins to aid in distribution analysis.

- **Pie chart**: A circular chart with triangular segments that shows data as a percentage of a whole.

- **Treemap**: A type of chart that shows different, related values in the form of rectangles nested together.

7.7.1.3 Tools for visualization
- **Tableau-One** is the most common visualization tool used by many companies. It provides the integration of databases, Teradata, my sQL,

and some other additional provision to add for more efficient visualization for business purposes. For a large dataset, constantly evolving dataset, it shows the best performance in terms of graphics and pictures, and also presents the best results in artificial intelligence, machine learning, and Big Data applications.

Advantages
- Best visualization capabilities.
- User friendly
- Excellent for diverse of data.
- Edibility of mobile apps

Disadvantage
- Expensive application
- Scheduling the report for a specific process is not available.

Dundas BI
Dundas BI provides the best customizable data visualization with various types of facilities to offer efficient data for better visualization.

Advantages
- Flexibility in the visualization process.
- It supports large data sets and various chart types.
- Additional features are available for data extracting and normalizing.

Disadvantages
- There is no option for future prediction.
- It supports up to 2D charts.

JupyteR
Its web-based tool allows for the best use of equations and numerical applications, as well as sharing and distribution. Statistical modeling, numerical simulation, interactive computing, and machine learning are all excellent uses for JupyteR.

Advantages
- Rapid prototyping
- Visually appealing results
- Facilitates easy sharing of data insights

Disadvantages
- Collaborative process is tough.

- At times code reviewing becomes complicated

Zoho reports
Zoho analytics is a wide-ranging data visualization tool that integrates the business process and its business intelligence methods to provide online report service. It mainly supports big data applications and healthcare reports.

Advantages
- Easy report creation and modification; inclusion of practical features like email scheduling and report sharing; ample data storage; prompt customer service.

Disadvantages
- The dashboard becomes confused when there are enormous volumes of data, therefore user training needs to be addressed.

Google charts
It is one of the best data visualization tools everyone can use and easily learn, to implement in any type of application. Providing storage and analysis facilities is an additional advantage of the data ensemble in the visualization process.

Advantages
- User-friendly platform

- Easy to integrate data

- Visually attractive data graphs

- Compatibility with Google products.

Disadvantages
- The export feature needs fine-tuning

- Inadequate demos of tools

- Lacks customization abilities

- Network connectivity required for visualization

Visual.ly
Visual.ly is one of the data visualization tools on the market, renowned for its impressive distribution network that illustrates project outcomes. Employing

a dedicated creative team for data visualization services, Visually streamlines the process of data import and outsourcing, even to third parties.

Advantages

- Top-class output quality

- Easy to produce superb graphics

- Several link opportunities

Disadvantages

- Few embedding options

- Showcases one point, not multiple points

- Limited scope

7.7.2 Applications of AI in data analytics

- Accelerates data preparation

- Automates insight generation

- Allows querying of data

- Automates report generation and dissemination

- Business process applications

- Clinical research

- Drug production

- Health resource tracking

- Supplier risk analysis

- Sentiment analysis

7.8 Conclusion

Data plays a crucial role in all industries, and without proper storage, maintenance, computing, and process, we cannot handle the data efficiently. Nowadays tons of data are generated in day-to-day applications and activities, so the efficient way of data handling is by using AI techniques, which is very helpful to manage the data in many ways like analysis, prediction, diagnostics, etc. In this chapter, we mentioned the importance of data, types of data, how data is computed in the earlier stage, and what may be the impact of AI in data analytics. In the modern era, how data visualization is important and what tools are helpful to achieve better visualization, and its pros and cons are also discussed.

References

[1] Cao, L. (2018). Data Science. *ACM Computing Surveys*, *50*(3), 1–42. https://doi.org/10.1145/3076253

[2] Dewett, T., & R Jones, G. (2001). The role of information technology in the organization: a review, model, and assessment. *Journal of Management, 27*(3), 313–346. https://doi.org/10.1016/S0149-2063(01)00094-0

[3] Endel, F., & Piringer, H. (2015). Data Wrangling: Making data useful again. *IFAC-PapersOnLine*, *48*(1), 111–112. https://doi.org/10.1016/j.ifacol.2015.05.197

[4] Hamad, F., Fakhuri, H., & Abdel Jabbar, S. (2020). Big Data Opportunities and Challenges for Analytics Strategies in Jordanian Academic Libraries. *New Review of Academic Librarianship*, *28*(1), 37–60. https://doi.org/10.1080/13614533.2020.1764071

[5] Hariri, R. H., Fredericks, E. M., & Bowers, K. M. (2019). Uncertainty in big data analytics: survey, opportunities, and challenges. *Journal of Big Data*, *6*(1). https://doi.org/10.1186/s40537-019-0206-3

[6] Heer, J., Hellerstein, J. M., & Kandel, S. (2018). Data Wrangling. *Encyclopedia of Big Data Technologies*, 1–8. https://doi.org/10.1007/978-3-319-63962-8_9-1

[7] Lu, Y. (2019). Artificial intelligence: a survey on evolution, models, applications and future trends. *Journal of Management Analytics*, *6*(1), 1–29. https://doi.org/10.1080/23270012.2019.1570365

[8] Madhukar, M. (2017). IBM's Watson Analytics for Health Care. *Cloud Computing Systems and Applications in Healthcare*, 117–134. https://doi.org/10.4018/978-1-5225-1002-4.ch007

[9] Mangalaraj, G. (2022). Big Data Analytics in Supply Chain Management: A Systematic Literature Review and Research Directions. *Big Data and Cognitive Computing*, *6*(1), 17. https://doi.org/10.3390/bdcc6010017

[10] Mazilu, L., Paton, N. W., Konstantinou, N., & Fernandes, A. A. (2020). Fairness in Data Wrangling. *2020 IEEE 21st International Conference on Information Reuse and Integration for Data Science (IRI)*. https://doi.org/10.1109/iri49571.2020.00056

[11] Raban, D. R., & Gordon, A. (2020). The evolution of data science and big data research: A bibliometric analysis. *Scientometrics*, *122*(3), 1563–1581. https://doi.org/10.1007/s11192-020-03371-2

[12] Saif, S., & Wazir, S. (2018). Performance Analysis of Big Data and Cloud Computing Techniques: A Survey. *Procedia Computer Science*, *132*, 118–127. https://doi.org/10.1016/j.procs.2018.05.172

[13] Shabbir, M. Q., & Gardezi, S. B. W. (2020). Application of big data analytics and organizational performance: the mediating role of knowledge management practices. *Journal of Big Data*, *7*(1). https://doi.org/10.1186/s40537-020-00317-6

[14] Taherdoost, H. (2022). Different Types of Data Analysis; Data Analysis Methods and Techniques in Research Projects. *International Journal of Academic Research in Management*, *9*(1), 1–9.

[15] Tsai, C. W., Lai, C. F., Chao, H. C., & Vasilakos, A. V. (2015). Big data analytics: a survey. *Journal of Big Data*, *2*(1). https://doi.org/10.1186/s40537-015-0030-3

[16] What is AI Analytics? (2022, May 24). Retrieved September 28, 2022, from https://www.anodot.com/learning-center/ai-analytics/

[17] What is Data Science: Lifecycle, Applications, Prerequisites and Tools. (2022, September 4). Retrieved September 28, 2022, from https://www.simplilearn.com/tutorials/data-science-tutorial/what-is-data-science.

[18] Pandimurugan, M. parvathi and A. jenila, "A survey of software testing in refactoring based software models," *International Conference on Nanoscience, Engineering and Technology (ICONSET 2011)*, Chennai, 2011, pp. 571-573, doi: 10.1109/ICONSET.2011.6168034.

[19] Chinnasamy, A., Sivakumar, B., Selvakumari, P., & Suresh, A. (2018). Minimum connected dominating set based RSU allocation for smart Cloud vehicles in VANET. Cluster Computing, 22(S5), 12795–12804.

[20] Selvakumari, P., Venkatesan, K.G.S. (2014) Vehicular communication using Fvmr technique. nternational Journal of Applied Engineering Research, 2014, 9(22), pp. 6133–6139.

[21] Chinnasamy, A., Prakash, S., Selvakumari, P. (2016). Wagon Next Point Routing Protocol (WNPRP) in VANET. Wireless Personal Communications, 90(2), pp. 875–887.

[22] Chinnasamy, A., Selvakumari, P., Pandimurugan, V. (2019). Vehicular adhoc network based location routing protocol for secured energy. International Journal of Engineering and Advanced Technology, 8(5 Special Issue 3), pp. 226–231.

[23] Vanita Sharon, A. and Saranya, G., 2018, November. Classification of Multi-Retinal Disease Based on Retinal Fundus Image Using Convolutional Neural Network. In International Conference On Computational Vision and Bio Inspired Computing (pp. 1009-1016). Springer, Cham.

[24] Thanuja, R. and Saranya, G., 2018, January. Identifying and correcting imbalanced labelled image for multi-label image annotation. In 2018 2nd International Conference on Inventive Systems and Control (ICISC) (pp. 849–854). IEEE.

[25] Sathish Kumar L, Prabu AV, Pandimurugan V, Rajasoundaran S, Malla PP, Routray S. A comparative experimental analysis and deep evaluation practices on human bone fracture detection using x-ray images. Concurrency Computat Pract Exper.2022; e7307. doi: 10.1002/cpe.7307.

[26] N. Arunkumar, V. Pandimurugan, M. S. Hema, H. Azath, S. Hariharasitaraman, M. Thilagaraj, Petchinathan Govindan, "A Versatile and Ubiquitous IoT-Based Smart Metabolic and Immune Monitoring System", *Computational Intelligence and Neuroscience*, vol. 2022, ArticleID 9441357, 11 pages, 2022. https://doi.org/10.1155/2022/9441357

[27] Ajantha Devi, V. (2021). Analysis, Modelling and Prediction of COVID-19 Outbreaks Using Machine Learning Algorithms. In *Intelligent Data Analysis for COVID-19 Pandemic* (pp. 331–345). Springer, Singapore.

[28] S. K. Lakshmanan, L. Shakkeera and V. Pandimurugan, "Efficient Auto key based Encryption and Decryption using GICK and GDCK methods," *2020 3rd International Conference on Intelligent Sustainable Systems (ICISS)*, Thoothukudi, India, 2020, pp. 1102–1106, doi: 10.1109/ICISS49785.2020.9316114.

[29] Devi, A., & Ahmed, H. (2021). COVID-19 Under Origin and Transmission: A Data-Driven Analysis for India and Bangladesh. In *Emerging Technologies for Battling Covid-19* (pp. 121–137). Springer, Cham.

[30] Ajantha Devi, V., & Nayyar, A. (2021). Evaluation of geotagging Twitter data using sentiment analysis during COVID-19. In *Proceedings of the Second International Conference on Information Management and Machine Intelligence* (pp. 601–608). Springer, Singapore.

8

Artificial Intelligence and Data Science in Various Domains

Arun Kumar Garov[1]*, A.K. Awasthi[2], and Minakshi Sharma[2]

[1]Department of Mathematics, School of Basic, Applied and Biosciences, RIMT University, Mandi Goindgarh, Punjab, 147301, India
[2]Department of Mathematics, School of Chemical Engineering & Physical Sciences, Lovely Professional University, Phagwara, Punjab, 144411, India

Email: arunkumar170@yahoo.com; dramitawasthi@gmail.com; minakshisohareya@gmail.com

Abstract

This chapter comprises various parts. The first part consists of the introduction of artificial intelligence and its types and its history. The second part concerns achieving artificial intelligence and its tools and working. The third part is based on the emergence of data science, and the concluding part consists of the correlation between artificial intelligence and its applications in various domains. This part consists of various fields and real-life examples where artificial intelligence can be implemented and used.

8.1 Artificial Intelligence

The concept of artificial intelligence came decades ago, and many authors like Alan Turing, John McCarthy, Stuart Russell, and Peter Norvig gave their concept as well as a different definition for artificial intelligence but from all, it is concluded that "Artificial intelligence is the science of creating intelligent machines that can work and behave like human beings." There has been much discussion about the applications of artificial intelligence in daily life like; work, mobility, healthcare, the economy, communication, and the dangers of this technology to human existence. Artificial Intelligence tries to mimic the human mind by putting logic, reasoning, and problem-solving capabilities into machines or computers. This chapter will uncover what are facts and fantasies about artificial intelligence.

Alan Turing described artificial intelligence as the systems or machines which can replace human acts. It combines robust and computer science to solve the problems like forecasting in various fields. John Macarty first coined the term artificial intelligence in the year 1956. People in the past have imagined and even documented machines like individuals with qualities like mobility and reasoning. Ancient drawings, sculptures, and books also describe these types of machines. The idea of this field came very early and is finely described by Aristotle in "The Politics." Many incidents from where the idea was generated.

a.   Replacement of human by automated tools

b.   Movement of shuttles TO.

c.   FRO Blinking of eyes.

8.2 Germination of Artificial Intelligence

Automatic scheduling

Artificial intelligence undertakes efforts to tackle a problem with the help of mathematical logic by machines. In artificial intelligence, the duck plays an important role in this; the mechanical duck can quack, flap its wings, paddle, drink water, eat, and digest grains.

In 1495, in the Medieval period Leonardo Da Vinci designed a sketch of a humanoid robot; to replace humans with robots. At that time, it was very hard to sure whether it would work, but today the technology has proven it and named it as artificial intelligence.

A book named "Leviathan" was published in 1651 by Thomas Hobbes, and in the introductory part of that book, the possibility of building artificial animals was documented; due to this, George Dyson refers to Hobbes as the father of artificial intelligence. In 1738, Jacques de Vaucanson displayed the duck as his masterpiece, and then in 1801, Joseph Marie Jacquard built a punch card-controlled self-operated loom. In 1900 Frank Baum came up with "The wonderful wizard of Oz." Rossum's Universal robots in which the scientists wanted to fit the power of God as in humans; then in 1948, Dr. W. Grey made two small robots known as tortoises named ELMER & ELSLE. They do not have any preprogramming languages. This was the turning point for the quest for artificial intelligence, and from here, the growth of artificial intelligence started. In 1980 the backpropagation algorithm started rapidly for artificial intelligence. Artificial intelligence is the ability of machines to seemingly think for themselves when the machine can perform the task in which humans can think and machine solve problems by themselves on their ability to learn to reason by practicing repeatedly. Artificial intelligence is the field of behaviors that we consider to be intelligent. When there may be anything of the human brain there, we look at our heart's behavior. This gives rise to a further quest about how the human actually managed to behave; intelligent first artificial intelligence undertakes efforts to tackle such problems with rigorous mathematical tools artificial intelligence gives rise to further questions about what exactly constitutes intelligent behavior, what it is to have a mind how humans managed to behave intelligently.

8.2.1 Types of artificial intelligence

Developing a machine or intelligent machine by John McCarthy with a combination of sciences and engineering is known as artificial intelligence. Artificial intelligence is categorized into two parts, namely, capabilities and functionality. Based on functionality, artificial intelligence is divided into four types, namely reactive machine, limited memory, theory of mind, and self-awareness. These are based on following capabilities:

Type of Artificial Intelligence Based on Capabilities		
Artificial Narrow intelligence (ANI)	Artificial General intelligence (AGI)	Artificial super intelligence (ASI)

8.2.1.1 Artificial narrow intelligence

It is also known as weak intelligence. This artificial intelligence deals with only specific tasks as it has only limited capabilities. It is known as the first stage and named it as machine learning.

The examples of artificial narrow intelligence include Google maps, Alexa, Siri, and many more. These applications are designed to perform only specific tasks like Google maps help in navigation only, whereas Alex and Siri are virtual assistant technologies.

8.2.1.2 Artificial general intelligence

This intelligence is also known as strong artificial intelligence. This type of artificial intelligence possesses capabilities similar to human beings and can also think and behave like them. At present, there are no machines that have the abilities like humans have. It is known as machine intelligence, which is the second stage of artificial intelligence. The successful working of artificial general intelligence requires the combination of much artificial narrow intelligence, which is still not practical in nature.

8.2.1.3 Artificial super intelligence

This is the third stage of artificial intelligence, known as machine consciousness, and it is considered the machine that is even smarter than human brains. Artificial superintelligence is the advanced version of artificial general intelligence as it is also till now theoretical so that artificial superintelligence will become a reality after a long time.

8.2.2 Fields that form artificial intelligence

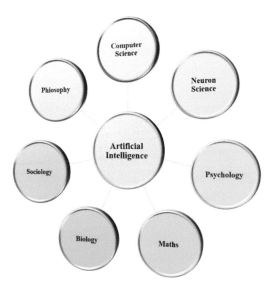

8.2.3 Is artificial intelligence a threat to human existence?

Researchers nowadays are more curious to know whether artificial intelligence is a threat, ban, or boon to human beings. The answer is based on these approaches, and the difference according to both approaches is given. Thus, it shows that it is a boon and ban depending upon the dependence of man on artificial intelligence.

Human approach	Ideal approach
• Systems having human-like thinking abilities	• Systems that think logically and wisely
• Systems that behave humanly	• Systems that act logical and wise

8.2.4 Branches of artificial intelligence

People often think that artificial intelligence, machine learning, and deep learning are the same, but it is not true that these are the subfields of artificial intelligence. Instead, its subfields include neural networks (NN), deep learning (DP), robotics, speech recognition, and natural language processing.

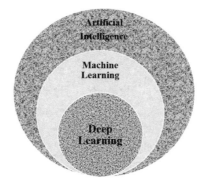

8.2.5 Applications of artificial intelligence

Artificial intelligence covers all areas. There is not any field where the arms of artificial intelligence have not spread. It covers all the fields as it is the need of hour. It fulfills today's demand; hence it covers the field of (i) medicine, (ii) education, (iii) robotics, (iv) information management, (v) biology, (vi) space, (vii) natural language processing, (viii) healthcare, (ix) business analytics, and (x) marketing.

It does not mean that only robots or above fields are using artificial intelligence, but the truth is artificial intelligence has found its way into our daily life. It becomes so general that we do not even realize that we are using it all the time like. Google how it correctly matches our search and shows exactly the same results. Facebook always gives content based on interest, and many more examples like this exist.

What has been discussed above is artificial intelligence totally works on different types of data, so there is a need to elaborate on types of data and their requirements in implementing artificial intelligence.

8.2.5.1 Definition of data required in AI

The data can be defined in many ways according to the requirement, but for AI's purpose, the suitable definition may be "The data is the information gathered in digital form which is used for reasoning, the calculation that can be transmitted or processed." The data is defined based on their storage capacities also. Nowadays, the data is stored in abundant forms, and a new unit is found for this, that is, brontobyte, which means 10 raises to the 27th power of bytes. The data which we get from the primary sources is raw data. In computers, the data stored is binary data which is in the form of either 1 or 0. The record of a specific quantity is data. It includes rows, columns, facts, and figures that can be used for survey and analysis. The data become information when arranged in an organized form. Moreover, for analysis, the authenticity of data depends on the source of data, that is, the primary source or secondary source of data.

8.2.5.2 Nature of the data

The nature of the data is classified into two types: qualitative and quantitative.

8.2.5.2.1 Qualitative data

Characteristics or attributes are shown in this type of data. It is for the descriptive type of data. The data cannot be computed or calculated but only can be described.

The data of the student samples have been taken, and the attributes are qualitative. It is also known as categorical data. For example, in the smartphone category, color and current phone rating are qualitative and categorized as qualitative data.

It is further subdivided into two parts: nominal and ordinal:

- **Nominal:** In this, the nominal data does not possess the natural ordering. The color of laptops is incomparable as it is not compared with others, so it is considered as nominal data type. Nominal data is not of quantifiable type, and hence it cannot be measured through numerical units. Genders like male and female are intangible, cannot be measured in any unit, and are regarded as nominal data types.

- **Ordinal:** This type of data maintains its class and also holds natural ordering. In the case of brands of laptops or smartphones, we can easily sort them in term of their size, that is, large > medium > small. The student's grading system is also an example of ordinal data where grading of A++ is better than grade A+, etc.These data types, not in numerical form, will be encoded into binary form so that the machine learning models can easily handle and perform various operations. There is one-hot and label encoding, similar to binary coding and integer encoding, respectively.

8.2.5.2.2 Quantitative data type

This data gives the numerical value as it quantifies things that make it countable. The price of anything, discount upon them, ratings of the product all these are of quantitative type. This type of data is not only for description or observation. It is for various calculations like forecasting, analysis, comparison, and many more.

For example, in schools, if we have taken the data of students who are interested in sports and it shows how many students like which game according to the data collected, this kind of information can be classified as quantitative and of numerical type.

The price of any product varies from *m* amount to *n* amount, and it can be further divided into fractional values so this type of data is also further subdivided or well explained in two categories:

- **Discrete:** This type of data has a fixed value in the form of integers or whole numbers. The number of research scholars, students, and films released this year are all examples of these types of data. Thus, from these examples, it is too said that these values cannot be measured but only can be counted as objects which have a fixed value. In this, the values counted are represented by hoe numbers, and it can also be identified through charts, bar graphs, tally charts, and pie charts.

- **Continuous:** In discrete values, there are no continuous values. Due to the whole values, it discontinued the values, which are the fractional values, as there is infinite number that lies between two whole numbers; hence the fractions that exist between whole numbers are considered continuous values. There are many examples of these types of continuous values like values of ECG and bandwidth of Wi-Fi.

The variation between the high value and low value is known as its range. As the temperature and weight of human break down, the continuous value can take any value. This type of continuous data is represented using a line graph for a certain period of time, frequency distribution, and histogram. It can be well used for analysis and interpretation.

8.2.6 Data collection

Data can be classified as primary data or secondary data, depending on its source of collection. Let us take a look at them both.

8.2.6.1 Primary data

The data collected directly without any intermediate stage is known as primary data. This type of data is collected and immediately used by a person for a specific purpose. Primary data is immaculate and original data without any loss of information. An example of primary data is the Census of India.

8.2.6.2 Secondary data

They are the data that one collects after one or many intermediate stages. A person collects this type of data and then passed on to some other person or organizations for some specific person, and therefore it is not original or direct data. This also means that this kind of data is adulterated or information is lost. This information is impure as statistical operations may have been performed on them already. This type of data is available on data repositories sites like WHO, Kaggle, and many government data websites.

8.2.7 Data science

As the name data science itself comprises of two words data and science, it is the systematic study of data. In this, it makes experiments with raw or structured data to get the information or knowledge. Today we can see that everything depends on data like business, education, entertainment, etc., so data is known as the driving component. Expert says that every second generates a quintillion of data from various sources like Facebook, Whatsapp, YouTube, Flipkart, Amazon, and many more websites.

Data play an important role in life through which we can make many decisions, and forecasting can be done from this data type. In this, it is observed that the type of data plays a significant role in drawing out the best knowledge and information. So it is mandatory to preprocess the data and make it error-free to get the desired result. The following are the most common types of data.

8.2.7.1 Big data can be divided into the following three categories

(i) Structured data, (ii) unstructured data, and (iii) semistructured data.

8.2.7.1.1 Structured data

The well-organized data, as the name shows, in structured form is known as structured data. The data that follows any pattern must be in tables or in another form that can be easily processed.

Data stored in spreadsheets and relational databases with multiple rows and columns are the best examples of structured data types.

8.2.7.1.2 Unstructured data

Data are neither structured nor organized, and as it is unorganized, so cannot be easily accessed or processed to draw out the information and knowledge.

Images and graphics types of data, along with many audio, video, email, and social media's data, are all unstructured data types. Apart from this, the locations sent by the people or their updation of exact location, and orders on Flipkart, Amazon, and Zomato are all examples of unstructured data.

8.2.7.1.3 Semistructured data

Data, that is in structured form but not in organized form, such as JSON, CSV, and XML types of files are the example of semistructured data files. To work on this, types of data and special software like Apache and Hadoop are required.

8.2.8 Data with artificial intelligence

Hundreds of applications are used daily, which are the implications of artificial intelligence (AI). For example, product recommendations on Amazon, movie recommendations on Netflix, voice assistant Alexa, filtering spam mail, and reality games like Pokémon are making use of artificial intelligence. There is always a creation of the latest trends in artificial intelligence. One of the interesting and useful trends is "unstructured data analysis using AI" which explains how artificial intelligence models are used to improve the analysis of large amounts of unstructured data. Big data and artificial intelligence come together to bring meaningful insights and help to make better decisions from unstructured data, which is very helpful for business development and growth.

Analyzing structured data using traditional analytical methods is simple, but almost 80–90% of data collected is unstructured data like text, video files, audio files, social media posts, etc., and these are difficult to analyze, but here, artificial intelligence does wonders. For example, Facebook uses deep face and deep texts, artificial intelligence models, for understanding sentiments and emotions, making them easier to analyze, and Twitter uses artificial intelligence that displays the most attention tweets. The built artificial intelligence models explore the data, identify the data types, and find relations between them, creating models that automatically speed up data to prepare tasks. This will help unstructured data to uncover ideas and discover new patterns that help in problem-solving and decision-making that extract ideas and patterns from large unstructured data. Some artificial intelligence models that are already in use are natural processing language, sentiment analysis, artificial intelligence bots, artificial intelligence forecasting, etc.

8.2.8.1 Structuring the unstructured data using artificial intelligence

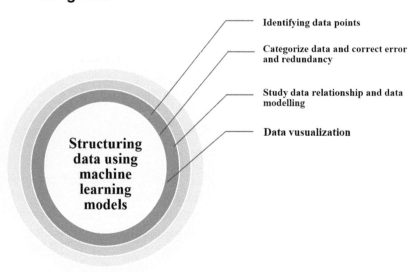

Now, it is easy for businesses to analyze billions of photographs, images, texts, and emails for their development. Therefore, unstructured data is made simple with artificial intelligence.

8.2.8.2 Steps to convert unstructured data to structured data for analysis and insights

Companies make use of unstructured data.

- Set goals for the organization.

- Create machine learning models to identify data points.

Prioritizing tasks that derive insights from unstructured data converting to structured data.

- Train the machine learning models.

- Remove errors and redundancies, if any.

- Proper data modeling is done.

- Testing the machine learning models.

- Get the required result, that is, structured data.

8.3 Workflows of Artificial Intelligence and Its Tools

8.3.1 Workflow driven by AI

AI models have received the majority of attention in modeling, with engineers who use machine learning and deep learning. They spent a large percentage of their time and energy developing AI models. Although modeling is an essential phase in the workflow, modeling is not enough or the end of the journey. Finding any problems early on and recognizing which parts of the workflow to concentrate time and resources on for the greatest outcomes are the keys to practical AI implementation success—yet these are not usually the most evident steps. The workflow of the AI has some steps given below;

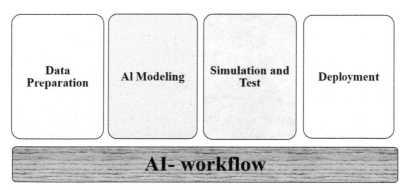

This four-step process makes up the AI-driven total workflow which can be understood in detail and see how crucial each one plays an integrating role of AI into a project.

- Step 1: Data preparation
  In AI workflow, the first and most important step is data preparation. In this, the neuron works for the nervous system in the same way data works. So, without accurate and well-prepared data, the system may not work properly or figure out why the model is not working. To find out the reason it may take number of hours.

  To train a model, one must have authenticated data, one of the most time-consuming data. Even engineers must focus on the preprocessing of data, modeling of data, and then implementation of data. There are many methods like amendment of parameters, fine-tuning, multiple iterations, etc.

  The importance of data preparation can be easily seen in construction and equipment companies, which are based on high volumes of data to form the working of various machines. Therefore, plenty of data is a must for accurate artificial intelligence modeling. Furthermore, the data can be split into training and testing types of data, so plenty of data is required to train and test the model. To make model easy, experts use MATLAB to clean or preprocess and label data for input in machine learning models.

- Step 2: AI modeling
  Artificial intelligence modeling is the second step. This stage of workflow works on the modeling of data which is used as input after the preprocessing stage. The rule for successful modeling is that it must create a tough, healthy, and fine model. To make such type of model, it mostly depends on data. Therefore, rather than thinking about deep learning or machine learning, most of the engineers used prebuilt models and made comparisons for the accuracy of models.

  MATLAB, Simulink are flexible tools for the iterative environment as it offers major support to the engineers. Engineers basically try to find out the various methods of using algorithms and also try to find examples regarding the modulation of algorithms and counter-examples for this. There are multiple domains on which the AI build models, and with the help of prebuilt models, engineers check and test the newly constructed models. Then, engineers make changes according to the desired goals by the number of iterations, and finally, the model is prepared in the form of algorithm.

- Step 3: Simulation and test
  Artificial intelligence, which was once a limited technology, today has applications in various fields, including robotic teachers and automated driving. AI models work within a larger system containing small AI and non-AI parts and must work with all these system parts for efficient working. Consider a scenario of an automated driving vehicle: Does the vehicle have any perception system for detecting objects (pedestrians, cars, and stop signs)? At the same time, such a system has to integrate with other systems for localization, path planning, controls, and more. Simulation and testing for accuracy are important factors to confirm whether the AI model is working properly within the system or whether everything sync well together with some other systems, and one has to confirm this before deploying a model into the real world.

For the highest level of accuracy and robustness, engineers always try to make a model so exact that it responds as engineers supposed.

To ensure the accuracy of the model, one must clear all the following questions:

- Is the accuracy of the model up to the mark as assigned?

- What is the model's performance?

- Does the model test for all the cases?

The model's authenticity and trust have been achieved, when it is successfully simulated and tested. Simulink is one of the tool by which engineers can verify the model's work.

- Step 4: Deployment
  Once the model is tested and found successfully built based on the checking from all the edges, then the model will be ready to deploy. In this, the next target is hardware and language. At this stage, it will be decided that in which language model should be deployed. In this step, engineers share an implementation-ready model. Moreover, the hardware environment also ranges from desktop to cloud to FPGA'S and MATLAB. In this, MATLAB can generate all the final codes in all scenarios. There are many platforms or environments to check the model's robustness in many ways rather than writing codes again.

When we deploy a model directly to the graphics processing unit, the coding errors are removed by the automatic code generation method, and it can also provide CUDA code that can be perfectly run on the graphics processing unit.

Artificial intelligence and its tools

Tools of the AI framework are used in several software for the analysis. Nowadays, machine learning (ML) is very active, or it can be said that it is the trend in the market. Therefore, AI and ML are available in every sector.

AI frame works Tools	
	Keras
	Scikit Learn
	Caffe
	MxNet
	PyTorch
	Auto ML
	Open NN
	H20: Open Source AI Platform
	Google ML Kit
	TensorFlow

8.3.2 Artificial Intelligence in data science

Today's time is known for its digitalization, which is both very hot topic in this digital age. Artificial Intelligence and data science use data for analysis and for various outcomes. These fields are correlated to each other in many ways. As development is the law of nature, in these fields, there are many advancements by which the fourth revolution came in the industry. Artificial Intelligence works on data with tools for analysis and develops permanent intelligent systems for future use in various fields.

8.3.3 Data science

In the multidisciplinary discipline of data science, information may be gleaned from organized and unstructured data using statistics, computer science, and machine learning methods. As a result, data science is one of the most lucrative, complicated, and rapidly expanding careers in the past 10 years. As a result of the realization of its importance, a plethora of data science companies have emerged to provide data-driven solutions across a range of sectors.

8.3.3.1 Process of data science

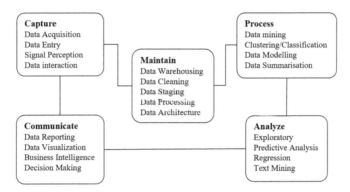

8.3.3.2 Life cycle of data science

There are fixed methods that are predefined for artificial intelligence to work in data science applications; these are known as the life cycle of data science.

The following methods are mentioned which are to follow in data science:

- **Capturing of data**: The first step is to collect the raw data. In this step, the data is to be refined, and it follows the extraction process. There are many sources like data entry, signal reception, etc.

- **Maintenance of data**: After cleaning the data, the data is stored in a data warehouse.

- **Processing of stored data in the warehouse**: The data from the data warehouse has been taken for further refinement by processing it and summarization.

- **Analyzation of data**: In this, various analyses are performed like Regression, forecasting, classification, etc.

- **Communication and visualization of result**: By implementing the various intelligence tools, the data report is given in the form of tables, graphs, etc.

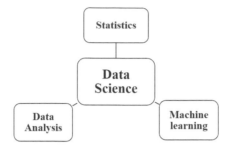

8.3.3.3 Applications of data science

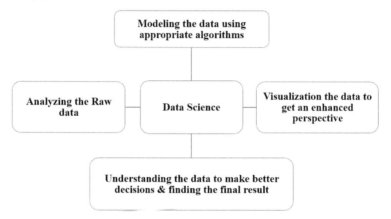

As the name indicates, data science works on data, and data is abundant as it is increasing quickly. Experts say that in every microsecond, a quintillion of data is generated. To analyze intelligent machines, data science in collaboration with artificial intelligence works prominently in various sectors like banking, manufacturing, transport, healthcare, E-commerce, etc.

8.3.3.4 Co-relation of artificial intelligence in data science

Artificial intelligence and data science are highly correlated to each other. These both have data as a common element. Data science works on data, and artificial intelligence works on processed data.

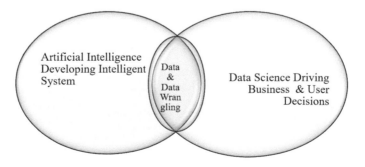

8.3.3.5 Role of artificial intelligence in data science

Artificial Intelligence plays an important role in enhancing the competencies of data science. These both can be interchangeably said as these both work simultaneously and depend on one another. Data science works on data, but artificial intelligence works on processed data. Data science is to find out or to know about its hidden patterns, but artificial intelligence is a tool by which data science completes its goal like forecasting, etc.

Artificial intelligence's main goal is to make machines independent so that without even man's intervention, machines can make the best decisions, solve problems, use logic, and reason out various causes and problems. In this, the professionals work on machines to develop new algorithms for future use. Keras, Spark, TensorFlow, Scala, Scikit Learn, etc., are tools used by artificial intelligence.

Data science is solving the problem statement, drawing a conclusion from the big data, and finding hidden patterns in the stored data. Data science uses artificial intelligence as a tool to do all these practices, such as Tableau, Python programming language, MATLAB, TensorFlow statistics, Natural language processing (NLP), and many more.

8.3.3.6 Trends in data science and artificial intelligence

With the passing of time, the trend of AI and data science is increasing. As the day dawns, and the role of AI and data science starts, people nowadays depend on machines for their work. The development in automation and advanced natural language processing shows increasing trends. Many technologies follow these like IoT, advanced analytics, UI, and cyber security. Due to these fields, human assistance is negligible, and machines work independently.

8.3.3.7 Future of data science and artificial intelligence

Healthcare, advertising, banking, education, and machinery are seeing the highest number of implementations of data science and artificial intelligence.

8.3.3.8 Domains of artificial intelligence and data science

8.3.3.8.1 Business

It empowers businesses by making good decisions. It helps the individual and warns about errors and malfunction before it completely breaks down. In business, it helps professionals in many ways.

- It enhances human and machine interactions through the user interface.

- It acts as a powerful tool to explore new markets and target populations for the business.

- It helps to analyze the business models and solve various related issues.

- It helps in solving production problems and makes the production competent.

- It helps in designing blueprints for business products and projects.

- It is used in customer service centers to render better customer service by efficiently solving customer problems or issues.

8.3.3.8.2 Healthcare

Dendral was the first AI problem-solving program developed for identifying bacteria and recommending antibiotics. The scope of AI in healthcare is widespread in various fields: health research, medical imaging, genetic research, personalized genetics, medical records, and many more.

Artificial intelligence technology is utilized in the development of various healthcare applications. These applications can run on mobile phones and help the user in the effective management of a healthy lifestyle.

There are many artificial intelligence healthcare applications like Healthyfy me, Google fit, etc.

Artificial Intelligence technology is used to develop devices and wearable devices that track day-to-day patterns and behavior of the human body. The devices based on artificial intelligence for tracking patterns, etc., are Smartwatch and Fitbit. Artificial intelligence can predict various diseases based on the provided data and helps patients take precautions and avoid the disease. In addition, artificial intelligence is being used in the form of robotics to perform operations in the operation theater.

8.3.3.8.3 Space

Indian space research organization (ISRO) has developed a national remote sensing center that processes satellite data and forecasts climate change and natural disasters. This technology uses optical remote sensing geographical data and artificial intelligence.

Artificial intelligence creates space-like conditions on Earth for imparting training to astronauts. It is also used in space crafts to develop voice assistant technology. This technology helps the astronauts in space to control various functions of spacecraft.

The robots that explore the moon and Mars surface are also developed by artificial intelligence and forecast the results with data science. One such robot, namely "Perseverance" was sent by Nasa to explore the surface of Mars.

8.3.3.8.4 Entertainment industry

Entertainment industry is a billion-dollar industry in India, with a value of rupees 596 billion in 2020. However, the rise of streaming platforms like Netflix and Amazon Prime Extra has taken this industry to the next level. The streaming platforms gather information about their consumer and then utilize a technique to detect the taste and liking of their customers and then recommend the content based on their findings.

VFX technology is very popular nowadays. However, this technology cannot be possible without artificial intelligence. The growth and success of animated and live-action movies have also witnessed new Heights in recent times; the success of these movies can be attributed to artificial intelligence.

Artificial intelligence technology is also used to review trailers and movies before their release.

8.3.3.8.5 Banking and finance

Banking and finance are one of the early industries to go for artificial intelligence. The banking and finance industry has a very large customer base which results in capturing and storing data from their customers. The effective utilization and processing of these data need some intelligent techniques. For example, artificial intelligence can be found in developing and implementing credit rating programs like CIBIL and CRIF. The rating agencies gather the borrower's credit history from different financial institutions and, based on that credit history, awards a credit rating or credit score to the borrower.

Artificial intelligence is effectively utilized in the development of banking apps. Some examples are SBI YONO1111111 and UCO M banking plus. Artificial intelligence is utilized in ATMs, also.

8.3.3.8.6 Automobile sector

The automobile industry has witnessed revolutionary changes in the last few years from traditional fossil fuel, and the industry is shifting its focus to alternate green energy sources like green hydrogen and electricity extra. This shift in the automobile industry is bringing artificial intelligence to the forefront. It is widely seen in navigation, voice control, automatic gear shift, and auto brake systems. In addition, the evolution of robotic and self-driven cars has further cemented the scope of artificial intelligence in the automobile sector.

8.3.3.9 Domains of artificial intelligence

The domain or targeted area of artificial intelligence is not bounded. It is classified into three main parts: formal tasks, mundane tasks, and export tasks.

8.3.3.9.1 Formal tasks

This type of task involves only formal logic. As in mathematics, proving the theorems and verification are formal tasks.

8.3.3.9.2 Mundane tasks

It is also known as ordinary tasks. In this type of task, fundamental operations are performed. The child learns perception, vision, and speech; all are mundane tasks.

8.3.3.9.3 Expert tasks

This type of task requires analytical and thinking skills. These types of tasks cannot be performed by ordinary people. It may be done only by professionals in manufacturing, design, planning, etc.

Computer vision, automated stock trading, recommendation engines, fraud detection, personalized online shopping, enhanced imaging and surveillance, video games, healthcare, and chatbots are the most common examples of real-world applications of artificial intelligence systems today.

8.3.3.9.4 Defense PSUs

The application of artificial intelligence has been finalized for each of the defense public sector undertakings (PSUs) and the roadmap of artificial intelligence (AI) has been discussed in Rajya Sabha. An AI roadmap has also been finalized for each defense PSU under which 70 defense-specific AI projects have been identified for development, and 40 projects have been completed. Under the chairmanship of Defence Minister Rajnath Singh, Defence Artificial Intelligence Council (DAIC) provided to necessary guidance and structural support for the application of AI to use for the military.

8.3.3.9.5 Healthcare

There are numerous patients and plenty of data, which is very difficult to handle, so artificial intelligence can forecast the patient's outcome and prevent them from various health issues. Data science can be helpful for the analysis of big data and for drawing out the kind information from it.

8.3.3.9.6 Industries

These two companies can drive the life cycle even without human intervention. Thus, in industries, it is a dire need for better results and outcomes.

These tools help to forecast the industry's loss and profit. Based on it, various industrialists act accordingly and can save everything before it becomes a reality. Most companies and industrialists are moving toward these fields for better services and products due to their efficiency, time-saving, and no human intervention.

8.4 Artificial Neural Networks

Neural networks run by a network will get higher cognitive metrics by adding one or more hidden units. It consists of only one layer, and in this layer; each layer has a specific number of neurons and hidden units. The input signal transfers the neural network from the inner layer to the outer level through a secret level in the feed-forward neural network. A neural network is the interconnection of computer processors like human brain neurons. An artificial neural network is simply known as a neural network, and it is of various types:

1. Recurrent neural network

2. Convolution neural network

3. Feed-forward neural network

4. Multilayer perceptron neural network

5. Perceptron neural network

6. Modular neural network

7. Radial basis function neural network

8. TensorFlow 2.0 neural network

9. Generative adversarial neural network

10. Deep belief neural network

11. Long–short-term memory neural network

The backpropagation algorithm is another name for the multilayer perceptron model, which can be completed into two stages, the forward stage and the backward stage. In the forward stage, the input layer works as activation, and the output layer as a termination function. Weight and bias values are modified as per the model's requirement in backward stage.

8.4.1 Algorithm of neural network (ANN)

Required Input:	OFS: Pass optimized feature set as input data to ANN. With the help of a genetic algorithm, input is implied to ANN. Let us see if there are a total of 500 rows in the satisfied category with eight attributes from the dataset. Its size will be 500*8, out of which if 60 rows are not selected, then 440*8 with group label 1 will be passed to the Neural network as an input. C: Target/category based on the field chosen. Class labels in the proposed case will be nonsatisfied, moderately satisfied, and satisfied. N: Number of neurons obtained.
Obtained Output:	Net: Trained structure (This structure will contain the propagated weights through neural training.)

1. **Start Forecasting**

2. **Load Training Data, P-data = OFS**

3. Start the basic parameters of ANN like N (Number of Neurons), Epochs, Performances, Technique used, and data division.

4. For I = 1 to T data

5. **If T-Data > High Category**

6. Group (1) = Features (OFS)

7. **Else if Medium >T-Data< High Category**

8. Group (2) = Features (OFS)

9. **Else belong to low Category**

10. Group (3) = Features (OFS)

11. **End – If**

12. **End – For**

13. **Initialize ANN** utilizing Training data and Group

14. **Net = Newff ()**

15. **Set the training parameters according to the requirements and train the system**

16. **Net = Train** (Net, T-Data, Group)

17. Return: Net as a trained structure

End – Function

8.4.2 TensorFlow 2.0

TensorFlow is the acquiring data of training model for solving forecasting and future results created by the Google brain team. TensorFlow is an open-source collection for numerical calculation and a comprehensive machine learning process. TensorFlow, together with the study of machine learning and deep learning models, constitutes artificial intelligence and algorithms, which make them useful by way of common metaphor who will use artificial learning. It uses artificial intelligence to transport products to improve the search engine.

8.4.2.1 Sessions in TensorFlow

A session in TensorFlow is used to run a computational graph to evaluate the nodes.

```
Import tensor flow as tf
x = tf.constant(10.0, name= 'x', dtype = tf.float32
y = tf.constant(20.0, name= 'y', dtype = tf.float32
z = tf.variable(tf.add(x,y))
init = tf.global_variables_initializer()
with tf.session() as session:
session.run(init)
print session.run(y)
```

8.4.2.2 Variables in TensorFlow

These are in-memory buffers that store tensors.

```
# Declare a 2 by 3 tensor populated by ones

v = tf.variable(tf.ones({2,3}, dtype=tf.float32)|
```

8.4.2.3 Constants in TensorFlow

In TensorFlow, constants are created using the function constant.

```
Constants in Tensor Flow
# syntax : constant(value, dtype=None, shape=None, name='const', verify_shape=False)

z = tf.constant (5.2, name = "x", dtype=tf.float32)
```

There are a few examples where it is applied.

1. Artificial intelligence can be easily seen in Google search engine, where we type the keyword, and the next word is automatically suggested in the search option.

2. The suggested videos, documents, and many more contents on social sites are based on the individual's interest due to artificial intelligence.

3.   It is being used by a lot of companies in the industries.

4.   Airbnb ingeniously uses the applications of TensorFlow to provide hospitality services and online platforms as a marketplace.

5.   TensorFlow is used in healthcare industries nowadays as healthcare trains network to discover particular anatomy.

6.   Fraud detection in PayPal is only possible by using TensorFlow.

7.   Deep transfer dream is a platform for generating modeling in computer science. In this, the complex fraud patterns detected for the improvement in the accuracy of products and to enhance the usage of experiences of workers.

8.   TensorFlow is used in mobiles made in China to improve its quality in the network issue.

9.   Cutover channel wire has been created, and the first time these wires are used TensorFlow can automatically forecast the cutover.

10. Time window verifies log operations and detects network anomalies.

11. Home subscriber server is one of the best successful systems of IoT where time window is already operated and working.

8.4.3 TensorFlow: features and applications

TensorFlow 2.0 supports eager execution by default. It allows building models and running them instantly. Python function uses IF function to mark JIT compilation. It runs as a single graph. Autograph feature of IF function helps to write graph code using natural Python syntax. TensorFlow offers multiple levels of abstraction but try to choose the right one to build and train models by using the high-level Kera's API that works and starts with answer flow. Machine learning becomes very easy if one or more flexibility Iker execution allows for immediate iteration and intuitive debugging (Figure 8.1).

When it is unable to eager execution, it will execute TensorFlow kernels immediately rather than constructing a graph. So now, it provides a direct path to production whether it is on a server's device or on the web TensorFlow.

Let us train and set up a model. The web TensorFlow trains the model without depending on the type of language to set up the model. These techniques set up the models and train them to make them flexible and provide firm performance. It also provides extra functions like API and subclasses of the models. Similar to other softwares, there also provides a powerful

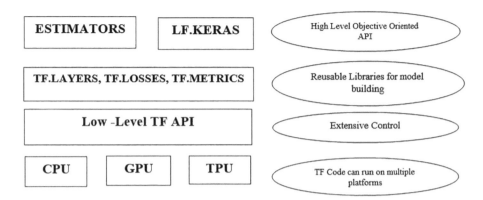

Figure 8.1 TensorFlow toolkits hierarchy.

platform of adds, libraries, and various models to do experiments on this floor, which is also directly imitative from its center. The framework does flow all the computations which involve tensors, so a tensor is a vector matrix of various dimensions that represents the type of all the operations which are conducted inside the graph, and the graph is a set of computations that express the successful age operation which is called as an open note.

Open notes are connected to each other. There is a floor for the developers to create a structured data flow graph that describes how the data moves through a graph and shows a series of processing notes that are not in the graph, representing a mathematical operation and connection. These notes are of a multidimensional data array. In this, the tensile sensor test floor provides the programmer with Python language, which is easy to learn and work with. It also provides a convenient way to express how high-level obstruction can be coupled. The tensor in the TensorFlow Python object has various applications. The math operations are not performed in Python, but the library has a transformation available through TensorFlow written in C++. It has high-performance binaries Python just between the pieces, provides high-level programming attraction, and builds a neural network together. Machine learning and deep learning process have single step, but, in this case, it is of low terms. It is so simple, and a typically have a machine learning license.

Ecolab has many processes, which also have some of the steps like the collection of data set, then building the model, training the network evaluating the model, and then forecasting the outcome in case of TensorFlow.

CNT the Microsoft cognitive tool, life sensor flow, uses a graph structure, but it focuses more on creating deep learning neural network which handles neural network jobs faster hand and has a broader set for Python, C++

Sea shark, but CNT is not currently easy to learn or deployed as TensorFlow. The Apache MX net Amazon deep learning framework can scale almost linearly multiple GPU and machines. It also supports a wide range of languages API like Python C++, JavaScript, and Julia, as, in a way, it is pleasant to work with TensorFlow. It is also in the market and other things because of various features, and it is the fact that it is open-source and has huge community support that not only provides and searches for a way to build new models but also forms a platform to interact with others.

In this, there is a construction of a construction phase, then execution phase where it builds a graph and evaluates the graph. Then based on this, it creates a session that initializes all the variables. The example of geometric sequences has been proved by TensorFlow 2.0, and it is so easy to execute. There is a TensorFlow 2.0, the latest released, that is even easier to code. It has eager execution by default, making things so much simpler and easier. Even professional can see strong codes, which are strong few lines of either execution or program.

8.4.4 Tensors

The atomic unit in TensorFlow is a tensor, a multidimensional array with uniform type. It contains elements of a single data type called a d type which is known as data type. Tensors are like NumPy errors. All tensors are immutable like Python numbers and strings. Tensor never allows updating the contents of a tensor, only can create a new one. Tensors are identified by the three parameters: rank, shape, and type.

Rank is the unit of dimensionality described with intensive. It is the number of dimensions of the tensor. In shape, the number of rows and columns together defines the tensor-type shape. In type, it describes a data type assigned to tensor elements. The tensor is of a zero-dimensional type known as a scalar, having no magnitude, and in this scalar has a zero axis, which is of rank 0. A one-dimensional or rank one tensor has only one row and column, known as a one-dimensional vector tensor. Density is similar to a one-dimensional vector. It has an axis, and it is of 1 rank, then a two-dimensional matrix. In the same way, it will be two-dimensional and of rank 2. Dimensional numbers will be defined as a table of numbers, continuing up to n-dimensional and n-axis (Figure 8.2).

8.4.5 Generative adversarial network

Adversarial network is an unsupervised learning and a deep learning-based rating model. This system is to compute neural networks to contend with

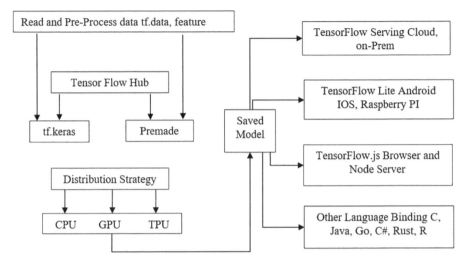

Figure 8.2 TensorFlow 2.0 architecture.

one another. E N GOOD FELLOW described this term in 2014 in a research paper 2014 then in 2016, Alex Radford proposed a model theory which is also known as DGCA or also known as voltage Dip for National general and virtual network. Most of Gangster De uses the general formulation and virtual networks. Generative models and Discriminating models are two well-known submodels of gain architecture. In this, it takes the sample data to generate the data that is either taken from the real sample or made by using a binary classification problem. Then, with the sigmoid function's help, it generates the output ranging from 0 to 1.

8.4.5.1 Working of GANN
Adverse data network in the generator means the unsupervised learning approach-based model. The adverse shall model is trained in an adverse serial setting and network in a generator model. It simply means the training of a model where it uses a neural network as artificial intelligence.

8.4.5.2 Algorithms in GAN
In this, the very first step is to generate a sample of data from the sample. It generates data that is cither taken from the real sample or made using binary classification. Then, with the sigmoid function's help, it generates an output ranging from 0 to 1. A generative model analyzes the distribution of data, and in this, the probability of discriminator maximizes the mistakes in the data, which is either taken from the real data or not. This is also done in the same way as it is in the hard-based model. In both models, the probability

distribution is estimated for both types of data and almost in the same way. In this, the generated process can be formalized in the mathematical formula where a generator is equal to discriminate on P-data access, the distribution of real data is P. Data that is the distributor of generator access in the sample from the real data Z is the samples from the generator RDX is the discriminator network in a generator network.

Training phase gives the real sample as from random noise samples are generated through Airtel network; the best examples, Go It Will Go, The Discovery network, which can detect the real and fake type of data; along with this, it also focuses on working of data.

8.4.5.3 Training phase

In training phase, generative adversarial network has actually two phases. In the first phase, train discriminator data and freeze the generator terms, which means it trains the data set and returns either fake or real value based on the data check as there is no backpropagation method. It only makes checks and classifies the discriminator that is either trained with real data or just classified correctly. In the second part, the generator can identify and classify into fake class likewise in the first step. Again, the generator and freezer works on the first phase result and try to make it better from the first phase and the whole process has been done to make it better so that it can be easily understandable in layman's terms also.

- Steps of training phase

    ○ Define the problem and collect the data.

    ○ Choose architecture of GAN.

- Depending on problem, GAN architecture can be chosen.

    ○ To Train the discriminator on real data.

    ○ To forecast them as real and number of times.

    ○ Generate fake inputs for generator.

    ○ Train the discriminator on fake data.

So whatever samples are generated from the generator network will going to train the discriminator to forecast the generated data as fake here we know that the discriminatory is actually forecasting the values as correctly, and then in last, it with the

- Train the generator with the output of discriminator.

So, after getting the simulator prediction with train the generator to fool the discriminator, we trained the GAN to actually get our solution from the problem, which is right defining the problem.

8.4.6 Applications of GANN

8.4.6.1 Forecasting of next frame picture

This can be used for various purposes including in which we cannot determine the activities in a frame that gets distorted due to other factors like rain, the small smoke extra. So, the possibility for people to know about the next frame in a video actually helps a lot in various activities. Based on this, people can generate full videos.

8.4.6.2 Text-to-image synthesis

The application of GANN can help to form text-to-image objects and attentive GAN, which is known as an object. This process can be performed in two steps. In the first step, generate the semantic layout, then generate it by synthesizing the image by using a deep convolution image generator which is the final step.

8.4.6.3 Story formulation

The layouts and refined details by synthesizing the words and another study about the GAN can forecast the whole story from a few words, so that is a very good idea. The application of GANN does not end here, and this can be used to give a few layouts and captions; based on that, it will generate images for the third application also.

8.4.6.4 Image-to-image translation

As a model which is designed for general purpose, that is, image-to-image translation forms a real image. It generates an image which is a fake image, and then it will be reconstructed to the previous image, which was a real image this is a mainstream and transition workplace that enhances the resolution of an image by an adverse neural network known as a gain in better quality. This is actually a very good GANN application.

8.4.6.5 Generation of super-resolution images from low-resolution images

There is an interactive image generation, and researchers have come up with a model that can synthesize a re-acted face animated by a person's movement while preserving the appearance of the face; at the same time, there are

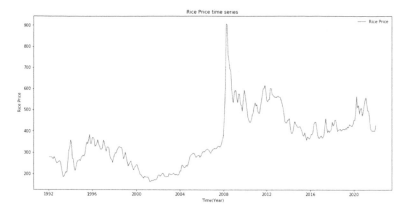

Figure 8.3   Rice price data from February 1992 to January 2022.

a lot of applications on which future researchers can work. In most cases, availability and safety issues yield redundant N+1, N+2, or 2N data center designs, seriously affecting power consumption.

8.5 Real Life-based Examples

The previous section discussed artificial intelligence and data science knowledge with its applications and with examples. The next section will discuss a case study to understand the forecasting analysis.

8.5.1 Case study 1

This case study is based on the data science analysis (big data analysis) concept of price time series data. Where used a machine learning technique for time series data analysis. For the time series, data analysis purposes use a time series data of rice crop, price data (in USD) is taken for the period February 1992 to January 2022 from the Index Mundi website. The dataset has recorded 360 values of rice price. The machine learning technique used Python 3 (Jupyter notebook 6.4.8). Figure 8.3 shows the time series of the rice price data in USD. In the rice price time series, it makes more sense to assume the sum of three terms: trends, seasonality, and residual. Figure 8.4 shows the decomposition of the rice price time series into three components such as trends, seasonality, and residuals.

This data science (big data) time series analysis can provide the ARIMA(p,q,d) model with the different order values of p, q, and d. The

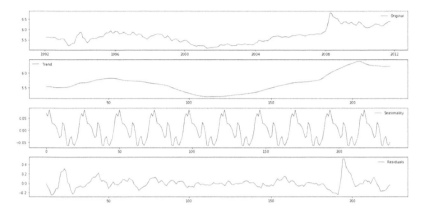

Figure 8.4 Decomposition of rice price time series into three components: trend, seasonality, and residuals.

analysis used the ARIMA model needed values of p and q, calculation of these values required a plotting of the autocorrelation function (ACF) and partial autocorrelation function (PACF), which helps to get the order values of AR and MA. Figure 8.5 shows ACF and PACF of time series at 40 lags.

8.5.2 Result and discussion

In the price time series big data analysis of rice crop using models;

1. ARIMA(1,1,1)

2. ARIMA(1,1,0)

3. Naïve forecast

ARIMA model applied for the analysis for the order of (1,1,1) and (1,1,0) in Python. Analysis of the model is shown in Figures 8.6 and 8.7, respectively.

Analysis of time series data by the model ARIMA(1,1,1) and ARIMA(1,1,0) of taken data values from February 1992 to January 2022, and for future calculated the forecast for the time series data, which is shown in Figures 8.8 and 8.9 for ARIMA(1,1,1) and ARIMA(1,1,0), respectively.

In Naïve forecast analysis of rice price time series data used programming processes in the Jupyter notebook are shown in Figure 8.10. Analysis

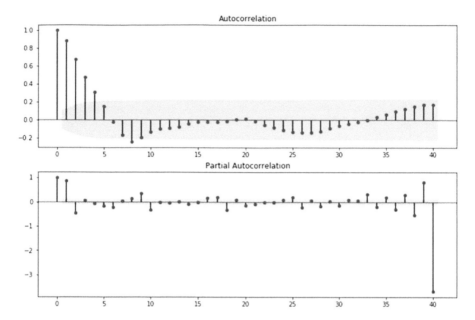

Figure 8.5 Autocorrelation function (ACF) and partial autocorrelation function (PACF) of rice price time series at 40 lags.

by the Naïve forecast model of price time series data is graphically shown in Figure 8.11.

This case study is based on big data analysis of rice price time series data and for that, used models are ARIMA(1,1,1), ARIMA(1,1,0), and Naïve forecast. Which helps to others to understand, how can apply models on data science as time series data. For clear concepts or knowledge, and ideas then do practice yourself.

8.6 Conclusion

Artificial intelligence is beneficial in every sphere of life when it is under human control. However, if it starts controlling humans, it will become a vice for society as a whole.

"It is useful when it is mastered by man; when it masters man, it becomes disastrous."

ARIMA Model Results

Dep. Variable:	D.Price	No. Observations:	359
Model:	ARIMA(1, 1, 1)	Log Likelihood	-1672.190
Method:	css-mle	S.D. of innovations	25.501
Date:	Thu, 17 Mar 2022	AIC	3352.380
Time:	00:37:16	BIC	3367.913
Sample:	03-01-1992	HQIC	3358.557
	- 01-01-2022		

	coef	std err	z	P>|z|	[0.025	0.975]
const	0.4410	1.869	0.236	0.814	-3.223	4.105
ar.L1.D.Price	-0.0260	0.137	-0.190	0.850	-0.294	0.242
ma.L1.D.Price	0.4260	0.125	3.407	0.001	0.181	0.671

Roots

	Real	Imaginary	Modulus	Frequency
AR.1	-38.5095	+0.0000j	38.5095	0.5000
MA.1	-2.3473	+0.0000j	2.3473	0.5000

Figure 8.6   ARIMA(1,1,1) model analysis of rice price time series data.

ARIMA Model Results

Dep. Variable:	D.Price	No. Observations:	359
Model:	ARIMA(1, 1, 0)	Log Likelihood	-1677.277
Method:	css-mle	S.D. of innovations	25.867
Date:	Thu, 17 Mar 2022	AIC	3360.553
Time:	00:37:22	BIC	3372.203
Sample:	03-01-1992	HQIC	3365.186
	- 01-01-2022		

	coef	std err	z	P>|z|	[0.025	0.975]
const	0.4497	2.054	0.219	0.827	-3.576	4.476
ar.L1.D.Price	0.3363	0.050	6.771	0.000	0.239	0.434

Roots

	Real	Imaginary	Modulus	Frequency
AR.1	2.9734	+0.0000j	2.9734	0.0000

Figure 8.7   ARIMA(1,1,0) model analysis of rice price time series data.

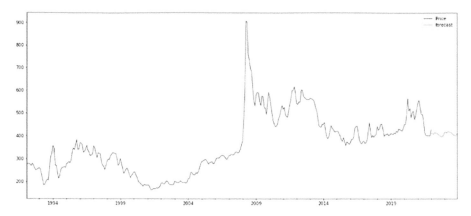

Figure 8.8   Rice price time series forecasted result by model ARIMA(1,1,1).

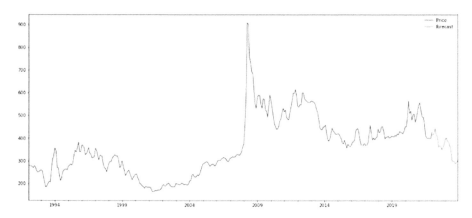

Figure 8.9   Rice price forecasted result by model ARIMA(1,1,0).

```
# predictions using naive approach for the validation set.
dd= np.asarray(train['Price'])
y_hat = valid.copy()
y_hat['Naive'] = dd[len(dd)-1]
plt.figure(figsize=(12,8))
plt.plot(train.index, train['Price'], label='Train')
plt.plot(valid.index,valid['Price'], label='Valid')
plt.plot(y_hat.index,y_hat['Naive'], label='Naive Forecast')
plt.legend(loc='best')
plt.title("Naive Forecast")
plt.show()
```

Figure 8.10   Programming processes of Naïve forecast model in Jupyter notebook.

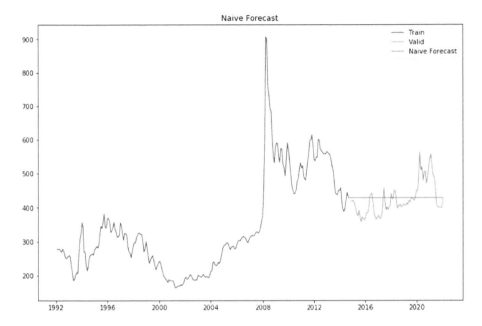

Figure 8.11   Analysis by the Naïve forecast model of price time series data.

Thus, in this, the machines have a type of general intelligence similar to humans, and in the coming years, humans will be replaced by machines.

References

[1] A.K. Awasthi, S. Kumar, A. K. Garov, 'IoT in the Healthcare Sector', in Machine Learning, Deep Learning, Big Data, and Internet of Things for Healthcare, Chapman and Hall/CRC, pp. 107–123, 2022.
[2] M. Sharma, 'Data Mining Prediction Techniques in Health Care Sector', In Journal of Physics: Conference Series, 2267(1), pp. 012157, IOP Publishing, May, 2022.
[3] S. Jindal, P. Saini, 'Internal and External Threat Analysis of Anonymized Dataset', In Handbook of Research on Intrusion Detection Systems, pp. 172–185, IGI Global, 2020.
[4] S. K. Sheoran, P. Yadav, 'Machine Learning based Optimization Scheme for Detection of Spam and Malware Propagation in Twitter',

International Journal of Advanced Computer Science and Applications, 12(2), 2021.

[5] A.K. Awasthi, M. Sharma, 'Comparative analysis of forecasting models in healthcare (COVID-19)', International Journal of Health Sciences, 6(S3), pp. 8649–8661, 2022.

[6] P. Mecenas, R. Travassos, M. Bastos, A. Carlos, D. Normando, 'Effects of temperature and humidity on the spread of COVID-19: A systematic review', PLoS one, 15(9) (2020).

[7] A.K. Awasthi, A. K. Garov, 'Agricultural modernization with forecasting stages and machine learning', In Smart Agriculture: in Emerging Pedagogies of Deep Learning, Machine Learning and Internet of Things, CRC Press, pp. 61–80, 2020.

[8] S. S. Patil, S. S. Pardeshi, A. D. Patange, R. Jegadeeshwaran, 'Deep learning algorithms for tool condition monitoring in milling: A review', In Journal of Physics: Conference Series, 1969(1), pp. 012039. IOP Publishing, July 2021.

[9] A. K. Garov, A.K. Awasthi, 'Case Study-Based Big Data and IoT in Healthcare', in Machine Learning, Deep Learning, Big Data, and Internet of Things for Healthcare, Chapman and Hall/CRC, pp. 13–35, 2022.

[10] S. Jindal, P. Saini, 'Internal and External Threat Analysis of Anonymized Dataset', In Handbook of Research on Intrusion Detection Systems, pp. 172–185, IGI Global, 2020.

[11] A. K. Garov, 'Quantity Based weights forecasting for TAIEX', In Journal of Physics: Conference Series, 2267(1), pp. 012151, IOP Publishing, May, 2022.

[12] Devi, A., & Nayyar, A. (2021). Perspectives on the definition of data visualization: a mapping study and discussion on coronavirus (COVID-19) dataset. In *Emerging Technologies for Battling Covid-19* (pp. 223–240). Springer, Cham.

[13] Ajantha Devi, V. (2021). Analysis, Modelling and Prediction of COVID-19 Outbreaks Using Machine Learning Algorithms. In *Intelligent Data Analysis for COVID-19 Pandemic* (pp. 331–345). Springer, Singapore.

9

Method for Implementing Time-Control Functions in Real-time Operating Systems

Wilver Auccahuasi[1], Oscar Linares[2], Karin Rojas[3], Edward Flores[4], Nicanor Benítes[5], Aly Auccahuasi[6], Milner Liendo[7], Julio Garcia-Rodriguez[8], Grisi Bernardo[9], Morayma Campos-Sobrino[10], Alonso Junco-Quijandria[10] and Ana Gonzales-Flores[10]

[1]Universidad Científica del Sur, Perú
[2]Universidad Continental, Perú
[3]Universidad Tecnológica del Perú, Perú
[4]Universidad Nacional Federico Villarreal, Perú
[5]Universidad Nacional Mayor de San Marcos, Perú
[6]Universidad de Ingeniería y Tecnología, Perú
[7]Escuela de Posgrado Newman, Perú
[8]Universidad Privada Peruano Alemana, Perú
[9]Universidad Cesar Vallejo, Perú
[10]Universidad Autónoma de Ica, Perú

Email: wauccahuasi@cientifica.edu.pe; olinares@continental.edu.pe; krojas@utp.edu.pe; eflores@unfv.edu.pe; nbenites@unmsm.edu.pe; aly.auccahuasi@utec.edu.pe; milnerdavid.liendo@epneuman.edu.pe; julio.garcia@upal.edu.pe; gbernardo@ucv.edu.pe; mcampos@autonomadeica.edu.pe; ajunco@autonomadeica.edu.pe; agonzales@autonomadeica.edu.pe

Abstract

Currently there are many applications that require to be executed in real time, as is the case of the various control systems. These applications have a particularity to run in a given time. With the use of resources, these tasks are grouped and classified by a priority mechanism, among other qualities of real-time applications. We can indicate that autonomous equipment uses programming based on real time; satellites and microsatellites also use them due to the need to be able to execute tasks at specific times and with a specific duration. In this area of applications, many solutions based on real-time programming arise. In this work, we implement a method to have a real-time system based on the Linux kernel, where you can make applications with exclusive use of the resources contained in the computer. The proposal is to configure an operating system based on Linux. We present, as a result, the procedures to be able to patch the Linux kernel, as well as the different configurations that have to be done; we also present model programs where we use the different approaches that allow us to develop real-time programming, the method presented can be implemented in different processor architectures, as well as to control different devices, with which the method is also scalable to solve more complex problems that require better control.

9.1 Introduction

Technological progress is allowing the development of new opportunities in areas where it was considered difficult to improve, one of them is the control of processes in real time, most of the different equipment and devices work with operating systems to run different applications, in this sense most of the processes are subject to the robustness and management of tasks and processes by the operating system, as well as the allocation of resources to different tasks, In a literature search we found works where they are responsible for optimizing the control of the different peripherals with the intention of being able to optimize the work of the operating system, we have works where it refers to the exclusive use of the RS232 communication port, with the intention of being able to exploit the maximum communication capabilities and to execute the transmission of data at maximum speed, we must indicate that the RS232 communication port works on the serial communication protocol and many of the devices that are currently in operation work with these communication ports [1]. However, the control issue is still a very important topic, and we find works where an alternative for the control of electrical energy consumption is analyzed, through the use of new communication architectures, as is the case of IoT technology, through which the consumption values of the different devices are sent, as well as a mechanism of emergency shutdown action is proposed, when these values pass certain thresholds, considered average values of use [2].

We found works referred to the development of medical equipment, which must have clear processes in order to obtain the safety of both the user and safety, so it is proposed systematization about good practices for the development of medical equipment, which has impacted the development processes of the equipment which has been based on the stage-gates process, where the systematization has divided the activities of the process into five phases with nine functional groups where each group works has its activity and tools used to perform its work, the study method was used to validate the systematization, so this model was presented to three companies to buy their process about the development of medical equipment so that the good practices were as a guide for development [4].

We found works referenced the development of an IRB system and expect the system developed to be used for the development of smart medical devices, accesses, fast IRB certification, and easy follow-ups, which has provided a mechanism for communication and cooperation between both researchers and manufacturers generated through an online conferencing system [5].

We found works referring to the design of medical equipment in order to change the types of approaches being increasingly energy efficient where the ECG is the most used. In the case of any equipment with lower energy consumption compared to traditional equipment is considered as more ecological or efficient equipment reducing the count so it is proposed to design and implement an efficient ECG machine using I/O standards LVDCI (low voltage digital control impedance), SSTL (Stub series terminated logic) and HSTL (high-speed transistor logic) [6].

We found works referred to telemedicine used to provide health services in remote locations for which a telecare system with GSM functionality was implemented with an architecture, processes, and procedures included in the development where the design science research (DSR) has been used during the development, for which each block has been designed separately and then integrated, at the end they have been compared with calibrated medical instruments where it was shown that the system worked as intended after calibration [7].

We found works referred to the development of catheter OS used in medical applications, for which a high precision remote control system is developed with the use of a master-slave system with which security is guaranteed with a simple micro-force sensor design, then the experiment on the operational simulation "in vitro" has been carried out, where the results indicated that the proposed catheter OS works perfectly to control the teleoperations which can improve the effectiveness, operability within an aneurysm with force feedback for intravascular neurosurgery [8].

We found works referred to consider the algorithmic design of complex electronic medical equipment with an element of the biotechnical system considered as a variant of the cyber-physical system which contains a biological object, electronic medical equipment, and a potential user, where the design of medical equipment is considered as complex according to the characteristics of the biotechnical system where each development of one piece carries an individual approach, where the analysis of the structures of these systems and their characteristics allows a systematization of the sequence of operations used for the creation after the tests it could be demonstrated that the design of the algorithm has improved during the design of biotechnical systems for therapeutic purposes [9].

In this paper, we develop a method to work with a real-time operating system based on the Linux kernel, called RTLinux, by which we will explain the procedures to patch the Linux kernel and configure the kernel to make applications that run on the new kernel that executes tasks in real time, we describe the steps and files to replicate and use the method.

9.2 Materials and Methods

The materials and methods presented in this work is related to indicate the necessary procedures to configure a system with a real-time system based on the Linux kernel with the RTLinux version, for which it is required to make certain procedures necessary to run programs based on real-time control related. The following is the block diagram of the methodological proposal, as shown in Figure 9.1.

9.2.1 Problems related to control

Industrial processes related to machine control are considered to be the most delicate due to a very important characteristic which is the processing time, both the duration time and the time allocated for the use of resources, one of the important tasks of the operating system is the control of execution times of all programs, the allocation of resources such as the time allocated and the designation of priorities. Therefore, when it is required to control critical resources and processes, we need an operating system that supports the management of certain tasks that may have a critical designation.

For these tasks, a desktop operating system is not very useful, which requires an operating system that works in real time, for which there are several options according to the resources and needs; one of the systems that can be implemented is related to being able to resort to a configuration of a real-time system, based on X86 architecture using a Linux kernel, in the next

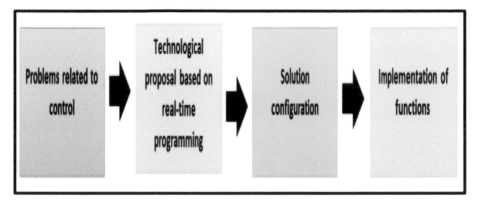

Figure 9.1 Block diagram of the methodological proposal.

pages we describe how to configure a real-time system from a conventional Linux system.

9.2.2 Technological proposal based on real-time programming

Technology offers us certain challenges which are important to face. In normality, we are working with fixed operating systems, where our work is related to the development of applications that run on the operating system. In this work, we let the operating system manage the execution times and the designation of resources. Faced with a need to control certain processes considered critical, we can resort to configuring a real-time operating system by patching a Linux operating system, this proposal can help in many cases without resorting to the purchase of a sophisticated operating system in the proposed method we work with equipment that we have available, and the task is related to have a real-time operating system based on the Linux kernel and RTLinux.

In Figure 9.2, we can visualize the difference between desktop operating systems and real-time operating systems, this comparison is important because it will allow us to evaluate the situations where it is necessary to work with real-time operating systems, in both cases it is necessary to have the hardware that is made up of all the resources necessary to execute the tasks, on top of the hardware runs the operating system, who are responsible for managing the execution of applications, depending on the tasks requested by the applications it is necessary to have these operating systems, An important feature for the choice of a real-time operating system is the critical level of the applications to be executed, in our case as the main theme is control,

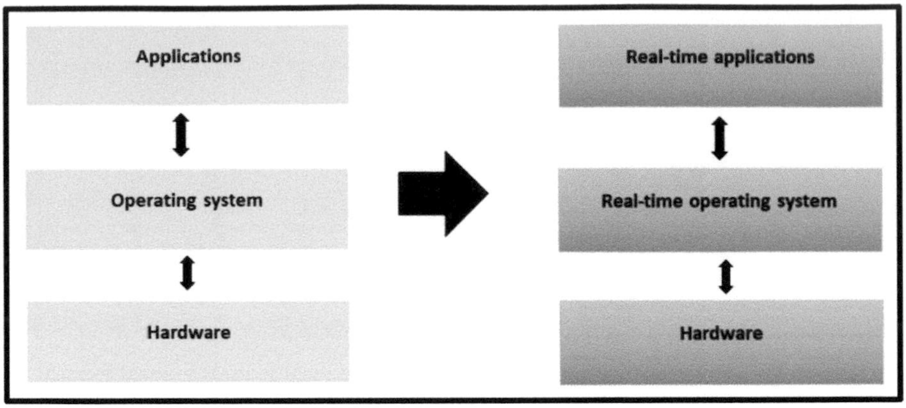

Figure 9.2   Comparison of a conventional and a real-time operating system.

it is necessary to have applications and systems that achieve the objectives of the applications, finally we have the top layer consisting of applications, these applications contain functions to fulfill the task entrusted and present the ways in which resources should be used as well as the time required for execution.

9.2.3 Solution configuration

The solution of the configuration is dedicated to describe the necessary steps to be able to pass from an operating system based on the Linux kernel toward a real-time system that executes a real time. Next, we present the necessary steps to be able to obtain a real-time operating system.

One of the first important tasks in the process of configuring an operating system in time is the decision to be able to know and decide which version and distribution to work with. In our case, as a demonstration, we will work with the version of Linux called RTLinux in version 3. 0, for which we need a version of RTLinux that is compatible with the version of the Linux kernel. For the demonstrative case, we need to have a version of the Linux kernel 2.4.4, and the version of RTLinux 3.0 must support version 2.4.4.4 of Linux. One of the first tasks is to download these files, and if you need to work with other versions of the Linux kernel, we must find compatible versions of RTLinux.

The next step is to create in the directory "/usr/SRC/" a folder with the name rtlinux-3.0. In this folder copies the rtlinux file. The procedure consists of decompressing the file, and we must consider that the downloaded file has the extension "tgz," so we must decompress the file. Then we execute

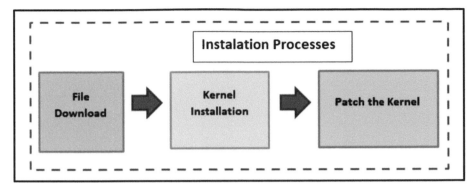

Figure 9.3 Block diagram of the installation process.

the command "make" from the directory. With this process, we have in the computer the first installation of the new kernel in real time.

At this point, we have two Linux kernel, one in desktop mode and the other in real time. The following procedure is to create an access link, commonly called patch it, for which we patch the "linux" file in the new folder created, this patch is important because it will allow access to the newly compiled kernel. It is important to indicate that at this moment in the computer, we have already compiled the two kernels.

Figure 9.3 shows the steps required for the initial installation of the real-time operating system kernel in the form of block diagrams.

Having installed the new kernel the next procedure is to configure the equipment to use the various resources of the computer. To access the configuration is resorted to the function "make menuconfig," in this menu, you can see all the resources that have the computer where it is being configured. In this part you select the computer resources with their specifications, as is the case of the processor model; in this configuration, you select the mode of use, a similar case for each of the components, as is the case of the hard disk drive, network card, memory, and peripherals. We finish this procedure by compiling again the new kernel with the "Build" command of the kernel, as shown in Figure 9.4.

Having configured the new kernel the next procedure is to update the boot system for which it is necessary to know which boot system is configured, which may be "LILO," at this stage, we update the boot configuration file. Then, finally, we have to configure the different modules that we want to work on, depending on the tasks and communication ports that work modules are downloaded and installed. If you work with data acquisition cards such as control units, it is necessary to install your module in order to use it.

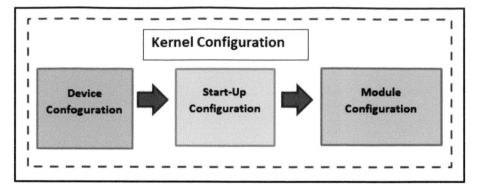

Figure 9.4   Configuration process block diagram.

Finally, we already have our system in real-time configured and ready to be used. When the computer starts, you must select the option that indicates RTLinux to be able to execute the operating system in real time. The next step is to carry out the programs having as directive the functions of real time, to be able to take advantage of the operating system in real time.

In Figure 9.5, you can see the architecture of a solution based on real time, which specifies some requirements related to control, which as implementation mechanisms, proceed to make applications to solve problems based on the control, which can provide a satisfactory solution, and to accomplish this task. It is necessary to have a real-time system, which is necessary to configure the system, as one of the main tasks in the control issue is related to the hardware [10] that controls them, you must configure the hardware to work with the new kernel in real time, so it is necessary that the various devices can be supported by the new kernel. In this way, we have an operating system in real time, configured and installed the hardware in the computer that solves the problems raised and the design requirements.

Figure 9.6, the flowchart of the configuration mechanism of the new kernel is presented. It is one of the most important because you have to configure the various devices and resources that the computer has, where the installation of the kernel is being performed in real time, the procedures described are related to the selection of the kernel to work as well as the support of the devices that are installed on the computer, it is necessary to indicate that many of these facilities are customized, The main objective is to solve the problem, the mode of use allows a computer to work exclusively with the equipment to be controlled, so it is important to have configured the necessary communication ports to solve the problem, it is recommended when you have the operating system working in real time, not to update,

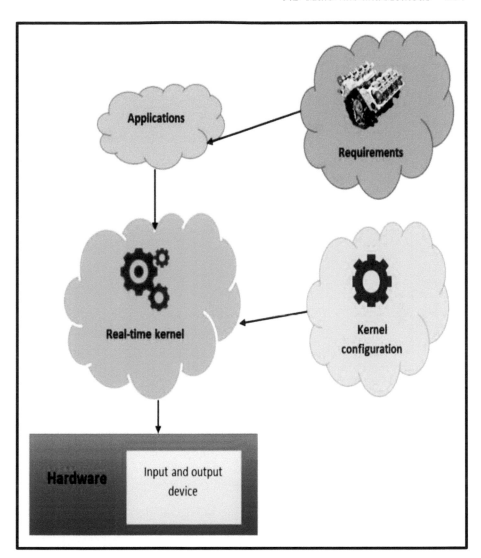

Figure 9.5 Architecture diagram for working in a real-time environment.

because the system is customized. A future update may lead to not support the configured devices. Therefore, we recommend leaving the new kernel locked to avoid future problems.

In the mode of use of computer equipment in industrial solutions, the computers are used exclusively with the equipment to be controlled, if necessary connected to the Internet, only if necessary. In most cases, these computers are not connected to the Internet. Another recommendation is not to place

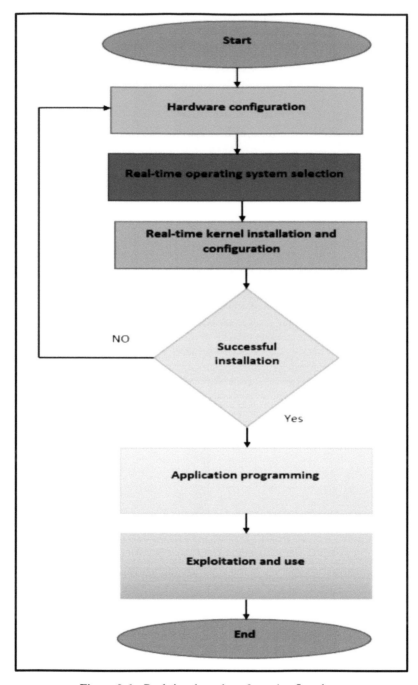

Figure 9.6 Real-time kernel configuration flowchart.

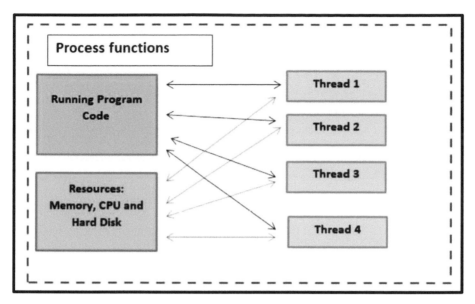

Figure 9.7 Diagram of the development mode of the functions in real time.

other devices or units because it may vary the communication port and thus spoil the configuration of the port that is working through control processes. New updates must ensure compatibility between the new kernel and the communication ports; in most cases, the useful life of the equipment is related to the computer equipment.

9.2.4 Implementation of functions

One of the fundamental tasks is that these programs are developed and compiled using the C language and its respective compiler. Furthermore, one of the important considerations is the programming mode, which works directly to control interruptions. The code indicates these tasks, indicating the resource address needed to solve the task. The applications have a particularity to control in a planned way all the available hardware, as is the case of the communication ports such as memory, enabling, and disabling, the use of memory is efficient because it works in static mode, working exclusively certain memory addresses for exclusive processes, with which you can manage the access times that are exclusive for specific applications.

Figure 9.7, we can see a typical configuration of the programming mode of the different functions; we must take into account that the generated code is executed in the new kernel in real time, and the working mode of

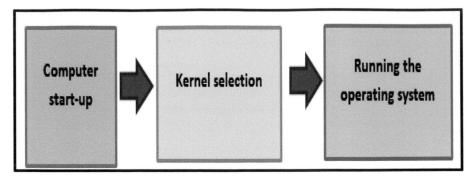

Figure 9.8   Real time kernel selection mode.

the functions indicates that the distribution of the functions is done based on tasks with which we can have from 1 to several tasks, in the case of the graph we indicate that we have four tasks, each of the tasks need the code represented by the function, as well as the resources that can be accessed, one of the main characteristics that can be seen in the graph, is that all tasks need the various resources, the way to solve this feature is developed based on a traffic light, where each task is given a priority, if a task arrives and requires some resources if the system detects that it has no available resource, it can finish or put to rest one of the tasks with lower priority. Therefore, when a programmer makes programs that must run in real-time operating systems, he must know the amount of memory and separate memory for each of the applications, this technique is known as static memory management, and although it limits access to memory, this method is adopted because the tasks in these systems are critical, so the risk is assumed, this is because normally all tasks have the same level of priority. In a practical way, the development is customized depending on the process to be controlled, defining all the necessary tasks.

9.3 Results

The results presented are related to the use of the operating system in real time, having performed the necessary procedures to install and configure the operating system in real time. When the computer is turned on, the system boot file is executed where it specifies the two operating systems, in order to differentiate, the new kernel must be renamed to differentiate it, with which we have two options, the first with the label "Linux," for a conventional Linux operating system and the other label "RTLinux," where the new kernel that identifies the operating system is executed in real time. Figure 9.8 shows the selection process of the new kernel.

Figure 9.8 shows the necessary steps to select the newly configured kernel. To have this selection option, the file that loads the system boot must be configured. To differentiate between both kernels, it is recommended to rename the real-time kernel as RTLinux.

Having configured the new kernel where the operating system performs tasks in real time, we have available on the same computer two kernels, each of them working in individual mode, accessing the normal kernel, we have an operating system based on conventional Linux, with access to all hardware resources and running the various applications, in this mode of execution-only runs the conventional kernel being the real-time kernel in off mode.

A second case is when we select the real-time kernel, where the kernel has access to all the hardware of the equipment, as well as the execution of the various applications. In this mode of work, the conventional kernel is in off mode, which shows that we can only run a single kernel under the same Linux platform, as can be seen in the graph in Figure 9.9.

In execution mode, we have several processes that can be running on the computer with the operating system in real time, which each of them is connected to the operating system through various libraries. That is why we need to install the modules to each hardware when configuring the new kernel in real time, these libraries communicate with the drivers for reading and writing in the available hardware resources, the management of resources are made based on the management of interrupts, thus having control of the resources at the exact time that the application requires it, as shown in Figure 9.10.

9.4 Conclusion

The conclusions we reached at the end of the presentation of the method are described from three aspects. The first is from a vision of the analysis of requirements in control needs. Demonstrated with a typical problem, in normal situations, when it is required to control mechanism, the choice of hardware is one of the main tasks; finding the best hardware ensures the success of the solution, in our case, to improve the performance of the solution, we also recommend the choice of a real-time operating system in addition to the selected hardware.

A second criterion is dedicated to configuring a real-time operating system based on a Linux platform. It is necessary to indicate that to choose this option, we must have an operating computer that can provide the solution to the problem, we must indicate that in this case, the real-time operating system will use the resources that are available on the computer, We must have

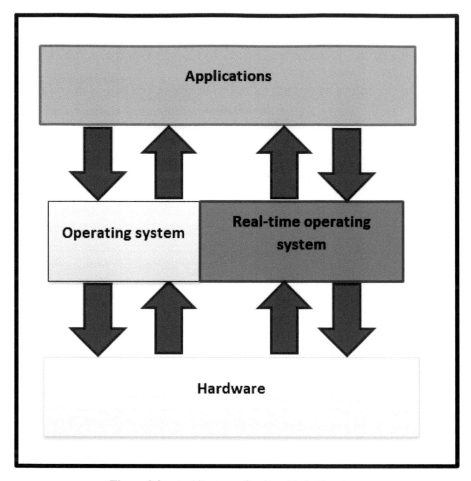

Figure 9.9 Architecture of a shared hybrid system.

in the computer two operating systems installed based on Linux, the first a normal kernel and the second running in real time, it is recommended to have a computer with the various communication ports necessary to be able to configure at the time of installation.

A third criterion is related to the mode of use in industrial applications. We must describe that these implementations are customized, so a solution can only run with specific hardware and equipment to control. The configurations are customized according to the needs. Having configured the solution, it is recommended not to upgrade because it can spoil the response of the systems. Normally, the new kernels require modern hardware. If they are installed with noncurrent hardware, the configuration may be compromised,

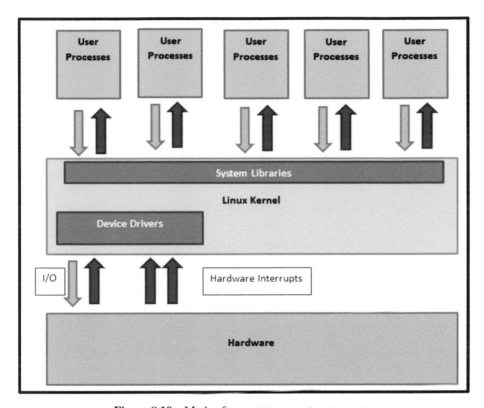

Figure 9.10 Mode of access to computer resources.

and the real-time system may not run normally, so it is recommended to lock the kernel and disable all upgrade mechanisms.

Finally, we can indicate that these solutions work exclusively with the equipment, not used for other applications, because it jeopardizes the config- uration of the communication ports, remaining as a protocol of use that the equipment is for exclusive use, the life of the equipment is linked to the life of the equipment with the operating system in real time, if in case an update is required, it is recommended to configure a new computer in real time as well as the integral configuration of the devices, before dismantling the previous equipment, any backup mechanism increases the probability of the continuity of the operation of the equipment.

The methodology can be applied and scaled in various applications. The usc will depend on the needs to be controlled as well as the possible solutions that can be given, having as main criteria the choice of the best hardware accompanied with a real-time operating system under a Linux environment.

Acknowledgments

We thank the authors who have made this work possible with the collaboration and contribution in each of the processes described in the development of the proposed methodology.

References

[1] Auccahuasia, W., Urbanob, K., Ovalle, C., Felippe, M., Pacheco, O., Bejar, M., ... & Ruiz, M. (2021). Application of serial communication in industrial automation processes.

[2] Auccahuasi, W., Urbano, K., Flores, E., Romero, L., Diaz, M., Felix, E., ... & Ruiz, M. (2022). Methodology to Ensure the Continuity of the Information Systems Service, Based on the Monitoring of Electrical Energy, Using IoT Technology. In Energy Conservation Solutions for Fog-Edge Computing Paradigms (pp. 283–307). Springer, Singapore.

[3] Aiquipa, W. A., Flores, E., Sernaque, F., Fuentes, A., Cueva, J., & Núñez, E. O. (2019, October). Integrated Low-Cost Platform for the Capture, Processing, Analysis and Control in Real Time of Signals and Images. In Proceedings of the 2019 2nd International Conference on Sensors, Signal and Image Processing (pp. 35–39).

[4] Lobato, K. C. D., Almeida, A. P. S. S., Almeida, R. M. A., Costa, J. M. H., & Mello, C. H. P. (2019). Good practices systematization for medical equipment development and certification process: A Brazilian case study. Health Policy and Technology, 8(3), 268–277. https://doi.org/10.1016/j.hlpt.2019.07.002

[5] K. C. Tseng and C. -L. Hsu, "IRB system for assisting the development of intelligent medical devices," 2009 IEEE International Conference on Industrial Engineering and Engineering Management, 2009, pp. 1022–1026, doi: 10.1109/IEEM.2009.5373512.

[6] T. Kumar, A. K. Memon, S. H. A. Musavi, F. Khan and R. Kumar, "FPGA based energy efficient ECG machine design using different IO standard," 2015 2nd International Conference on Computing for Sustainable Global Development (INDIACom), 2015, pp. 1541–1545.

[7] D. Tshali, M. Sumbwanyambe and T. S. Hlalelef, "Towards a GSM enabled remote monitoring medical system," 2018 3rd Biennial South African Biomedical Engineering Conference (SAIBMEC), 2018, pp. 1–4, doi: 10.1109/SAIBMEC.2018.8363195.

[8] S. Guo, H. Kondo, J. Wang, J. Guo and T. Tamiya, "A New Catheter Operating System for Medical Applications," 2007 IEEE/ICME

International Conference on Complex Medical Engineering, 2007, pp. 82–86, doi: 10.1109/ICCME.2007.4381697.

[9] L. Berezko, S. Sokolov, I. Yurchak and O. Berezko, "Algorithmic Approach to Design of New Medical Equipment," 2021 IEEE XVIIth International Conference on the Perspective Technologies and Methods in MEMS Design (MEMSTECH), 2021, pp. 12–15, doi: 10.1109/MEMSTECH53091.2021.9468093.

[10] R. P. Kumar and V. A. Devi, "Real time monitoring and controlling of Marine Sewage Treatment Plant Effluent," *OCEANS 2022 - Chennai*, 2022, pp. 1–6, doi: 10.1109/OCEANSChennai45887.2022.9775422.

10

Efficient Blurred and Deblurred Image Classification using Machine Learning Approach

E. Udayakumar[1], R. Gowrishankar[1], P. Vetrivelan[2], V. Ajantha Devi[3], and Seema Rawat[4]

[1]KIT-Kalaignarkarunanidhi Institute of Technology, India
[2]PSG Institute of Technology and Applied Research, India
[3]AP3 Solutions, India
[4]Amity University Tashkent, Uzbekistan

Email: udayakumar.sujith@gmail.com; rgowri2000@gmail.com; vetrivelan@psgitech.ac.in; ap3solutionsresearch@gmail.com; srawat1@amity.edu

Abstract

Images can be deteriorated for a variety of reasons. For instance, blurry images are produced by out-of-focus optics, while noise is produced by variations in electrical imaging systems. The blurred image classification and deblurring approach provided here uses DWT. After classifying the blurry image, there are ways to deblurring the given blurry image. The aim of blur image classification is finding blurred or unblurred images from input images. The deblurring of the photographs is presented toward the conclusion. This suggested deblurred image categorization and deblurred image can produce the best results. Finally, we evaluate the parameter analysis. This proposed scheme uses textural feature-based image classification using a neural network using a machine learning approach. Texture features are removed using the Gray level co-occurrence matrix, and the artificial neural network is advanced for the classification of images into dissimilar classes.

10.1 Introduction

Images of the corrupted model lack sufficient information. From one perspective, whether we know the level of cloudiness or not, deblurring procedures can be categorized as either "informed" or "blind." However, obtaining permission to use this material for different purposes should not come as a surprise. In informed deblurring, prior knowledge of the amount of distortion and its characteristics is required, while in blind deblurring, we assume that the distortion is unknown. In most cases, the Point Spread Function cannot be acquired. Current methods for blind deblurring typically describe the distortion process across the entire image [11].

Picture dark is a critical wellspring of picture degradation, and deblurring has been a celebrated assessment point in the ground of image planning. Many reasons can cause the picture to become dark, for instance, the ecological roughness (Gaussian cloudiness), relative camera development during the presentation (development dark), and point of convergence anomalies (defocus dark). The reconstructing of clouded photographs, that is, picture deblurring is the path toward prompting torpid sharp as a single model in uniform. Prior to the nonastonish deconvolution stage, which is the usual procedure for outwardly obstructed deblurring, the spot evaluation is carried out. Increasing picture priors in the greatest a posteriori evaluation is one of the conventional externally hindered deconvolution procedures. Priors such as edge priors and sparsity priors arc frequently viewed as being in the composition. In most cases, these algorithms employ an expectation maximization (EM) phase, which energizes the evaluation of the cloudiness section at one step and the sharp picture at the next development [13]. There are deserts that continue to use their programs, although image previous based approaches may appropriately evaluate the components similar to the idle sharp photos.

The main flaw of sparsity priors is their limited ability to cover small areas. When the image satisfied is homogeneous, the edge prior techniques, generally reliant on the image content, will simply flop. This work suggests a "learned prior" for the cloudiness piece evaluation that considers the Fourier change. The subject of no edges in a portion of the common picture patches is addressed by the repeat space feature and important plans. Notwithstanding how the information is fix based, our framework can manage greater picture patches that appeared differently in relation to sparsity priors. For example, locally (nonuniform) or universally (uniform) clouded. There are different advances engaged with perceiving the feeling of people introduced in a picture. These are face discovery and picture procurement, picture improvement, picture rebuilding, which are applied sequentially [14].

Presently a-days, the transformation of foggy pictures to sharp pictures is one of the primary issues since everybody utilizes advanced cell and snap pictures all over. Once in a while, the photos are caught as obscure in light of the camera movement or the snapshot of the item. The picture taker diminishes the screen speed for pleasing pictures below low light. When we decline the shade speed, the picture gets obscured on most occasions. Just as the picture gets obscured when the inclination of the picture taker's hand shakes, because of this issue, the change from obscure picture to sharp picture has become a habitually utilized proposal in picture preparation. To sharpen this kind of obscure pictures, the vast majority of the attempts have done [20]. Generally, these arrangements are executed utilizing profound learning calculations. In profound learning, explorers have been changing the foggy pictures over to sharp pictures in ease and productive ways, which permits them to be actualized in a wide reach. A picture contains terribly helpful data; once in a while, this data is needed for criminological labs, courts, etc. Thus, in the greater part of the news channels and papers, data is procured as a picture by passing on the data effectively to the individuals.

When innovation has likewise evolved by utilizing these advances, ordinary pictures have come instead of top-notch pictures. An ever-increasing number of advancements are made to simplify present-day life, including photograph-altering procedures. As common individuals procure the capacity to redesign, make, or convert advanced pictures, the validity of the pictures they work proof is turning into an issue [1]. Besides, as advanced pictures overwhelm our day-by-day lives, we might want to get a handle on whether a given picture is prepared, how well the picture is prepared, ought to get supplanted. To upgrade the image, a large portion of the procedures (or) strategies are accessible. By utilizing these techniques, most people want to drive the photos.

After applying these techniques, the image becomes clearer by changing the edges and setting the surfaces just as some necessary subtleties. In sign handling, varying the haze picture to a sharp picture normally decreases the issue of deconvolution, and the part execution of movement obscure [2] needs to continue as before. Notwithstanding, the general exhibition of the bit is frequently unaltered. In real cases, the plane turn of the camera or the moving item is obscured. The image is debased by two or three haze bits, alluded to as a topsy-turvy obscure picture. The magnificent technique is utilized to improve a picture, which implies that it can totally reestablish a sharp picture of various goals in a pyramid bit by bit. This is particularly valid for both streamlining-based techniques and, all the more, as of late neural organization-based strategies. Movement obscure in genuine photography is confounded to such an extent that the idea of move flimsiness is not continually present. It is exceptionally hard to get rid of the nonuniform second haze in the picture. The reasons for the nonuniform movement obscure might be isolated into two classifications: camera turn and target development [3].

To adapt to the difficulty of target development, nonuniform haze picture upgrade procedures have been suggested. Current approaches start with an entirely hard size of the obscured image and dynamically recover the image in high goal until the full goal is extended. Using this, we suggest a more effective organizational structure for staggered picture deblurring, and the method is effective. We search for a workable strategy to advance picture improvement, motivated by the example of the ongoing achievement of the encoder-decoder organization for several PC creative and perceptive assignments, the encoder-decoder blocks [6]. We demonstrate that applying a current encoder–decoder structure immediately does not currently produce the majority of worthwhile results. Using a different approach, our network proposal satisfies the requirements of various CNN designs and achieves preparation plausibility. Additionally, it produces improved results that are significant for huge movement upgrade [10].

10.2 Related Works

A neural network with convolutional layers organized by channels with substantial sizes should effectively replicate the reverse channels' fit for an input-obscured image. Due to the nature and size of these layers, an appropriate approach is needed to enhance the representation of the reverse channels. To achieve this, deblurring layers are introduced, defining a reversal process for a set of obscuring channels intended to capture the range of effects observed in input images. The reversal can be performed by recovering k inverses. However, the authors opted to use the mathematically more robust reversal process described in [22]. Furthermore, spatial channels are decomposed into a combination of two 1D channels. A series of convolutional layers with small-size channels is employed to harness the crucial information shared among themselves, forming the foundation of the denoising submodel. Addressing the need for suitable deblurring channels and constructing an extensive neural network architecture are key challenges in this concept. The model presented in this study proposes a modified version of this system that offers practical solutions to these challenges. This section begins by showcasing the employed model, followed closely by the selection of the planning cost function. Subsequently, the planning approach and the data recovery process are detailed [16].

There are various ways to handle and think about this problem. Deblurring approaches can be categorized as either astonish or nonblind from one point of view, depending on whether the dimness bit is known. While in externally debilitated deblurring we anticipate that the clouding director will be dark, nonshock deblurring demands the prior knowledge of the fog location and its boundaries. The point spread function (PSF) is not detected overall in sober interest; therefore, the application extent of the externally crippled deblurring [9] is much more commonplace than that of the nonamaze deblurring. In real applications, the only information we often need to manage is a single darkened image. The dimness digit of the entire image is typically represented as an in-the-stun deblurring method that are now in use. The segment evaluation is done before the nonblind deconvolution step, which is the standard procedure of outwardly impeded deblurring. One kind of the old style amaze deconvolution procedures incorporate improving picture priors in the most maximum a posteriori (MAP) evaluation. The two-stage plan for dark sort gathering and limit ID. Zeroing in on reasonable fog appraisal, we try to manage two difficulties in this paper. One of them is outwardly impeded murkiness limit appraisal from a lone (either locally or all around the globe) darkened picture without doing any deblurring.

The rebuilding of obscured pictures and photos, that is, picture deblurring is the way toward getting concealed sharp pictures from deficient data in the debasement model. In this manner, to eliminate obscure in pictures where the obscuring boundary is obscure and is privately obscured, a three-stage system is acquainted with characterizing the haze type, gauge the boundaries, and along these lines, deblur the picture. First, a pretrained dataset utilizing DNN (deep neural network) is initiated to recognize the haze type and GRNN (general regression neural network) to appraise the obscuring boundaries of each kind of obscure. Characterization and deblurring of pictures have numerous continuous applications, for example, in reconnaissance frameworks, satellite imaging, and wrongdoing examinations. A rotating minimization plot is embraced in this deblurring, denoizing, and deconvolution is essentially nonblind, which implies it needs earlier information on the bit and its boundaries [18]. Dazzle deconvolution mostly centers around improving picture, edge, and sparsity priors and utilizations a desire boost calculation. In this cycle, picture priors confine their applications, and sparsity priors just speak to little neighborhoods, which is a significant disadvantage. In this paper, camera shake is assessed by dissecting edges in the picture, viably by developing the Radon change of the piece edge priors that come up short if picture content is homogeneous, which is a significant disadvantage [5].

In picture incomplete haze location and order, obscure sort characterization is finished utilizing Naive Bayes Classifier for hand-made haze highlights, which is not profoundly productive or vigorous, and their pertinence is low on regular images [7]. In this paper, the issue of no-reference quality evaluation for computerized pictures undermined with obscure is tended to. A huge genuine picture information base is produced, and abstract tests are directed at them to create the ground truth. In light of this ground truth, various top-notch pictures are chosen and falsely debased with various powers of recreated obscure (Gaussian and direct motion). Then, the presentation of best-in-class procedures for no-reference obscure measurement in various obscuring situations is assessed widely, and a worldview for obscure assessment is proposed. These joined educated highlights work in a way that is better than hand-created highlights. Yet, these are prepared on irregularly initialized loads which could sometimes yield a helpless neighborhood ideal which is significant imperfection. It is a learning-based haze finder that uses joined highlights for the neural organization, yet it is not enthusiastically suggested as it does not generally ensure ideal outcomes [19]. The bilinear deep belief network (BDBN), a revolutionary profound learning model for analyzing interactive media content, is proposed in this research for picture characterization. By drawing on the engineering of the human visual framework and the method of perceptive discernment, BDBN intends to provide human-like judgment. Unique deep engineering with greedy layer-wise reproduction and global adjustment is proposed for the picture information structure. The deep model performs marvelously when indicated photos are absent since a BDBN is developed under a semi-managed learning method to correct certifiable picture arrangement tasks. It is a proficient DBN working framework that can likewise be consolidated for obscure grouping and boundary assessment utilizing machine learning [4].

This paper is worried about the computerized assessment of the recurrence reaction of a two-dimensional spatially invariant straight framework through which a picture has been passed and obscured. For the instances of constant direct camera movement and an out-of-center focal point framework it is indicated that the force cepstrum of the picture contains adequate data to recognize the haze. Strategies for deblurring are introduced, including rebuilding the picture's thickness adaptation. The rebuilding strategy burns through just a humble measure of calculation time. The noticed obscured patches utilized as preparing and testing tests have qualities that are not as clear as their recurrence coefficients which are not absolutely solid [12]. Picture obscure piece order and boundary assessment are basic for dazzle picture deblurring. This study suggests a two-stage architecture called Temporal

Deep Belief Networks (TDBN) to first group the haze kind and then identify its borders [31].

In the haze type grouping, this strategy endeavors to recognize the haze type from the blended contribution of different hazy spots with various boundaries. In order to extend the information tests in a space with discriminative elements and then aggregate those highlights, a semi-regulated DBN is ready. Additionally, the proposed edge recognition on logarithm range in the boundary distinguishing proof enables DBN to identify the hazy boundaries precisely. Here, as the recognizable boundary proof is additionally done utilizing the TDBN, it is not profoundly productive as the variety between obscure boundaries of a similar haze type is not as extraordinary as that between obscure kinds [20].

10.3 System Design

The existing strategy used here was Fourier change and SVM classifier. Fourier change is known as the repeat territory depiction of the main sign. The term Fourier changes implies both the repeat space depiction and the mathematical action that relates the repeat zone depiction to a segment of time [15]. The Fourier change is not limited to components of time; anyway, to have a united language, the space of the primary limit is routinely insinuated as the time region. Similarly, as in the examination of real ponders showing common apportionment (e.g., scattering).

Automatic modulation classification (MC) is a middle advance between signal discovery and demodulation and assumes a vital part in different regular citizen and military applications. It is likewise one of many key advancements in programming radio and intellectual radio. The acknowledgment techniques in the early years are essentially about sign waveform, recurrence, transient sufficiency, and transient stage. The exhibitions of these techniques slip immediately when they face low SNR. Measurable choice and example acknowledgment dependent on insights are two primary techniques for moving toward the MC issue. A few analysts use support vector machines (SVM) to care for MC issues, and get higher arrangement precision. Be that as it may, the two references neither offered how to choose the ideal boundary of an SVM classifier nor how to build multi-class SVM. In this paper, we initially present the 14 help vector machine, then research the choice techniques for piece capacity and its boundary, study the multi-classes grouping strategies, and apply them to computerized signal arrangement [21].

An efficient blurred and deblurred image classification is detailed in the following sections.

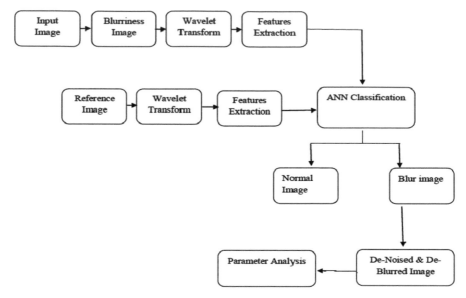

Figure 10.1   Block diagram of the proposed system.

10.3.1 DWT and NN classifier

See Figure 10.1.

10.3.2 Image denoizing

Image denoizing is a significant picture preparation task, both as a cycle and as a segment in different cycles. This channel is truly adept at safeguarding edges, yet easily differing areas in the info picture are changed into piece-wise consistent locales in the yield picture. Utilizing the TV channel as a denoize prompt tackling a second request nonlinear PDE [33]. Since smooth locales are changed into piecewise steady districts when utilizing the TV channel, it is alluring to make a model for which easily differing areas are changed into easily shifting districts, but the edges are safeguarded. Here, the furthest left picture is the first picture, the center picture is forced with commotion, and the furthest right picture is the reestablished [17] picture utilizing the fourth request model. Another methodology is to join a second and fourth request technique. The thought here is that the fourth request sifts smooth locales conspire, while edges are separated by a second request plot [23].

10.3.3 The image deblurring problem

Digital image is made out of picture components called pixels. Every pixel is allotted a force, intended to describe the shade of a little rectangular fragment of the scene. A little picture regularly has around 2562 = 65536 pixels, while a high-goal picture frequently has 5–10 million pixels. Some obscuring consistently emerges in the account of a computerized picture, since it is unavoidable that scene data "pours out" to neighboring pixels [25].

10.3.4 Artificial neural network classifier

Neural associations are a computational approach that relies upon a gigantic collection of neural units unreservedly showing the way in which a characteristic brain handles issues with enormous lots of natural neurons related by axons. Neural associations commonly include different layers or a strong shape plan, and the signed path crosses from front to back. Back inducing is where the forward instigation is used to reset loads on the "front" neural units, and this is at times done in blend in with planning where the correct result is known. The advantages of the proposed structure are that it offers a better hint to uproar extent and jam the image edges and surfaces, our system misuses the two pictures to make a phenomenal changed picture, reduced time use measure and no exceptional gear is required [27].

Imaging frameworks experience the ill effects of a few kinds of picture debasements. Movement obscure is a typical wonder, as are defocus obscure and climatic choppiness obscure. Every one of them lessens the picture quality altogether. In this way, it is basic to create techniques for recuperating approximated inert pictures from hazy ones to build the presentation of imaging frameworks. Such techniques will discover wide pertinence in different fields. Nonetheless, the issue of obscure evacuation is a famously badly characterized opposite issue that has puzzled researchers for quite a long time. An obscure can be displayed dependent on the defocus range. Therefore, it is more advantageous and pragmatic to tackle a boundary ID issue than to assess the PSF. It utilized the most extreme probability assessor (MLE) on the whole dataset accessible to appraise the boundaries for a Gaussian model [30]; see Figure 10.2.

The autonomous segment research non-Gaussianity measures to evaluate the bounds for complex models. To measure movement lengths, Run and Majhi proposed an extensive premise work neural organization with picture highlights based on the strength of Fourier coefficients. Yan and Shao offered

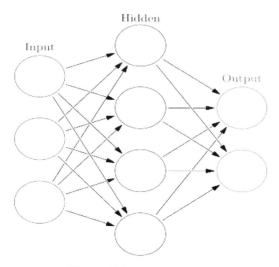

Figure 10.2   ANN diagram.

a managed profound neural organization to organize the hazy type. They were given the standard fix log probability technique to restore the inactive picture. The second was also used by Kumar et al. to evaluate the Gaussian change for choppiness obscurations [32]. Utilizing opposing channels or another nonblind deconvolution technique, the dormant picture can be restored. A nonblur arrangement is consistently produced by simultaneous evaluation of the PSF and the dormant picture, which is a nonconvex problem. A doable visually impaired picture deblurring structure is to gauge the PSF and the inactive picture then again. In any case, this structure requires earlier information on both the picture and the PSF.

The picture edges should be reproduced during every cycle before the subsequent stage. Moreover, a coarse-to-fine method should likewise be utilized to evade nearby minima during the exchanging improvement measure. All the disservices referenced previously make the deblurring method tedious. Besides, most existing methodologies are intended for arbitrary hand-drawn hazy spots. Nonetheless, in true circumstances, the haze model would consistently be known. On account of target identification on a transport line, the PSF, as it were, relies upon the movement length [34]. From this perspective, the PSF assessment system in the current techniques can be disentangled, and the speed of the deblurring calculation can be expanded. The images are obscured because of numerous reasons, for example, flaws in catching pictures, low power during camera introduction, air issues, etc. Yet imaging, like some other perception measure, is rarely great: vulnerability creeps into the estimations, happening as commotion and different corruptions in the recorded pictures. The picture is a projection of this present reality onto the lower dimensional imaging medium, a strategy that naturally disposes of data. Once in a while, the data lost may contain things we are keen on: it very well may be helpful to attempt to recuperate these concealed subtleties [24].

The current execution of the RadonMAP calculation requires a lot of memory. To tackle a one uber pixel picture, it needs around 3-4GB of RAM [8]. Noise evacuation, frequently called movement deblurring or dazzle deconvolution, is tested from two viewpoints. The main test assesses obscure bits or point-spread capacities (PSF) from obscured pictures. Since numerous clamor picture sets can clarify the noticed loud picture, obscure portion assessment is troublesome. Clamor assessment can be particularly troublesome if the commotion is spatial variation, for example, due to a unique scene or a camera turn. The subsequent test eliminates the clamor to recuperate a commotion-free picture. The clamor midpoints neighboring pixels and weakens high recurrence data of the scene [26]. Therefore, the issue of recuperating a commotion-free picture is not well presented, and it should be tended to by deblurring frameworks or calculations. Obscuring in pictures emerges from various sources, such as barometrical disperse, focal point defocus, optical abnormality, and spatial and worldly sensor combination. Human visual frameworks are acceptable at seeing it. However, the component of this preparation is not totally perceived. In this manner, concocting measurements to gauge obscure pictures is hard. Movement obscure is one of the striking wellsprings of corruption in photos. Although movement obscure can some of the time be alluring for aesthetic purposes, it frequently seriously restricts the picture quality [28].

Obscure relics result from relative movement between a camera and a scene during the introduction. While obscure can be decreased utilizing a quicker shade speed, this accompanies an unavoidable compromise with expanded commotion. One wellspring of a movement obscure is camera shake. At the point when a camera moves during the presentation, it obscures the caught picture as per its direction. We can alleviate the camera shake obscure utilizing mechanical picture adjustment equipment. In any case, when a camera takes a long introduction shot of a dim scene and a camera employs a zooming focal point, the camera shake can be excessively huge for assistive gadgets to oblige. Another wellspring of obscure is the development of items in the scene, and this sort of obscure is harder to stay away from. Along these lines, it is regularly attractive to eliminate obscure computationally [29].

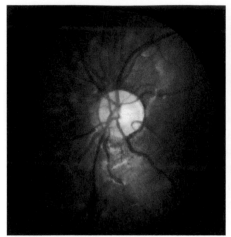
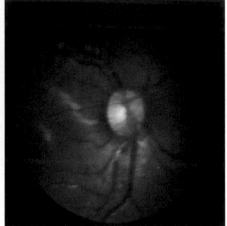

Figure 10.3   Input image.   **Figure 10.4**   Blur image.

10.4 Results and Discussion

See Figures 10.3 and 10.4.

Detection of blur image

Figure 10.5 shows that when we give an input image, it will change into different parts as DWT images that may be classified into the form of texture, shape, picture contrast, etc., and it automatically classified into four types of deblurred images. If the given image is blurring, then it will be trained by using ANN classifier and finally detect the image as "blur image" using machine learning (Figures 10.6 and 10.7).

Detection of normal image

Figure 10.8 shows that the given input image is a normal image, and then it will be followed by another image, a DWT image which could be the given input image in the form of four different images: input image, horizontal image, vertical image, and shaded image. Based on the images, it will be trained using an ANN classifier to detect the image as a "normal image."

Command window for blur and normal image

Finally, the values for two different input images are displayed in the command window as shown in Figures 10.9 and 10.10, where further details about the input image and segmented image regarding the shape, color, etc., are also provided.

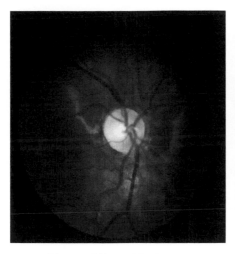

Figure 10.5 Deblur image.

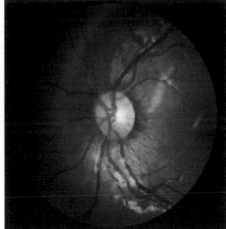

Figure 10.6 Normal image.

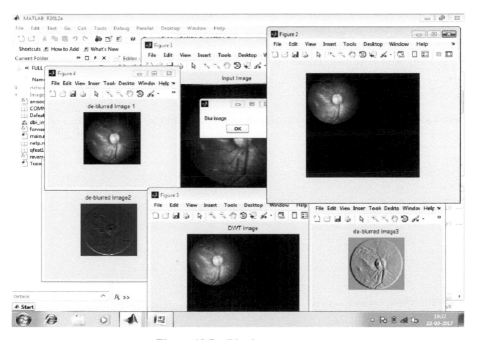

Figure 10.7 Blur image output.

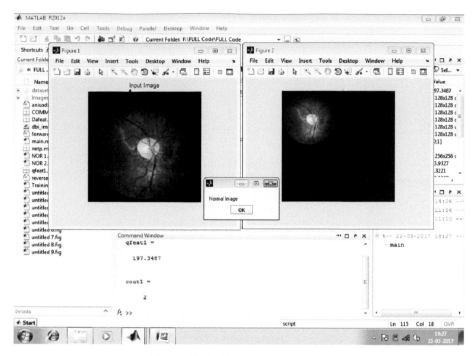

Figure 10.8 Normal image output.

Figure 10.9 Command window for a normal image.

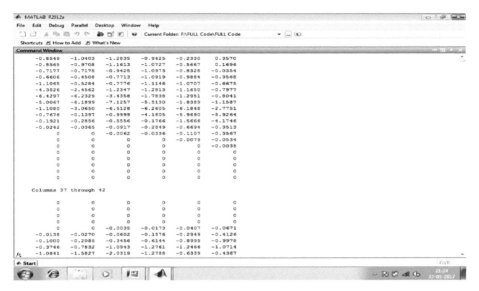

Figure 10.10 Command window for blur image.

10.5 Conclusion

The findings of the blurred and normal image are one of the merging difficulties in digital image processing. Numerous approaches have been projected to address this subject. One of the greatest issues in these strategies needed to bargain this was, having the option to identify the obscured and ordinary picture areas without being influenced by the normal picture handling activity, for example, pressure, commotion expansion, and turn. The other test was computational time, which becomes significant thinking about the enormous information bases, this procedure would be utilized on, and here the multi-wavelet DWT is being proposed to detect the images better than the existing method. And by using the DWT, the performance and the disadvantages in the DCT are eliminated using machine learning.

10.6 Future Work

Furthermore, for better future evaluation, the method DWT and DCT are combined during processing. Nevertheless, the approaches' performance was adequate. In the end, it is necessary to build quicker detection methods that satisfy the requirements for robustness. As a future work, the image detection process could be involved in moving images that are fast-moving images in the video streams. And later on, it can be processed using a single image to detect the classification of blurred and deblurred regions.

References

[1] Almeida. M and Almeida. L, "Blind and semi-blind deblurring of natural images," IEEE Transactions on Image Processing, vol. 19, no. 1, pp. 36–52, Aug. 2010.

[2] S.Yogadinesh and et al. "Certain Investigation of Identify the New Rules and Accuracy using SVM algorithm", Middle-East Journal of Scientific Research, IDOSI Publications, 23, pp, 2074–2080, 2015.

[3] Cho. T.S, Paris. S, Horn. B, and Freeman. W, "Blur kernel estimation using the radon transform," in Proc. IEEE Conf. Computer Vision and Pattern Recognition, Colorado Springs, CO, USA, 2011.

[4] Dr. S.S anthi and et al., "Region Growing Image Segmentation for Newborn Brain MRI", BioTechnology: An Indian Journal, Trade ScienceInc Journals, vol 12, Issue 12, pp.1-8, Dec 2016,

[5] Erhan. D, Manzagol. P, Bengio. Y, Bengio. S, and Vincent. P, "The difficulty of training deep architectures and the effect of unsupervised pre-training," in Proc. IEEE Int. Conf. Artificial, Intelligence, and Statistics, Florida, USA, 2009.

[6] S. Santhi and et al., "TB screen based SVM & CBC technique", Current Pediatric Research, volume. 21, pp.338-342, 2017.

[7] P. Vetrivelan and A. Kandaswamy, "Neural Network Classifier based Blurred and De-Blurred Image Classification for Bio Medical Application", Journal of Web Engineering, vol 17, issue 6, pp. 3858 – 3864, 2018.

[8] Joshi. N, Szeliski. R, and Kriegman. D, "Psf estimation using sharp edge prediction," in Proc. IEEE Conf. Computer Vision and Pattern Recognition, Anchorage, AL, USA, 2008.

[9] Krishnan. D, Tay. T, and Fergus. R, "Blind deconvolution using a normalised sparsity measure," in Proc. IEEE Conf. Computer Vision and Pattern Recognition, Colorado Springs, CO, USA, 2011.

[10] P.Vetrivelan and et al., "An Investigation of Bayes Algorithm and Neural Networks for identifying the Breast Cancer", Indian Journal of Medical and Paediatric Oncology, Medknow Publications, vol 38, Issue 3, pp. 340–344, 2017.

[11] Ajantha Devi, V., & Nayyar, A. (2021). Fusion of deep learning and image processing techniques for breast cancer diagnosis. In *Deep learning for cancer diagnosis* (pp. 1–25). Springer, Singapore.

[12] J. Rama and et al., "Automatic Detection of Diabetic Retinopathy through Optic Disc using Morphological Methods", Asian Journal of Pharmaceutical and Clinical Research, I Volume 10, pp. 28–31, 2017.

[13] P. Vetrivelan and et al., "An Identify of efficient vessel feature of Endoscopic Analysis", Research Journal of Pharmacy & Technology, Volume. 10, pp. 2633–2636, 2017.

[14] S. Sindhumathy and et al., "Analysis of Magnetic Resonance Image Segmentation using spatial fuzzy clustering algorithm", Journal of Global Pharma Technology, vol. 10(12), 88–94, 2018.

[15] C. Ramesh and et al., "Detection and Segmentation of Optic Disc in Fundus Images", International Journal of Current Pharmaceutical Research, vol 10(5), pp. 20–24, 2018.

[16] K. Yogeshwaran and et al. "An Efficient Tissue Segmentation of Neonatal Brain Magnetic Resonance Imaging", Research Journal of Pharmacy and Technology, vol 12(6), pp. 2963–2966, 2019.

[17] C. Ramesh and et al., "A Review on diagnosis of Malignant Melanoma from Benign Lesion by using BPNN and ABCD Rule Parameters", International Research Journal of Pharmacy, vol.9(10), 2018.

[18] Levin. A, Weiss. Y, Durand. F, and Freeman. W, "Efficient marginal likelihood optimization in blind deconvolution," in Proc. IEEE Conf. Computer Vision and Pattern Recognition, Colorado Springs, CO, USA, 2011.

[19] Levin. A, Weiss. Y, Durand. F, and Freeman. W, "Understanding and evaluating blind deconvolution Algorithms," in Proc. IEEE Conf. Computer Vision and Pattern Recognition, Miami Beach, FL, USA, 2009.

[20] Liu. R, Li. Z, and Jia. J, "Image partial blur detection and classification," in Proc. IEEE Conf. Computer Vision and Pattern Recognition, Anchorage, AL, USA, 2008.

[21] Shan. J, Jia. J, and Agarwala. A, "High-quality motion deblurring from a single image," ACM Trans. Graph., vol. 27, no. 3, pp. 721–730, Aug. 2008.

[22] Rugna. J and Konik. H, "Automatic blur detection formet a data extraction in content-based retrieval context," in Proc. SPIE Internet Imaging V, San Diego, CA, USA, 2003.

[23] Devi, V. A. (2021). Cardiac multimodal image registration using machine learning techniques. In *Image Processing for Automated Diagnosis of Cardiac Diseases* (pp. 21–33). Academic Press.

[24] Su. B, Lu. S, and Tan. C, "Blurred image region detection and classification," in Proc. ACM Multimedia, Scottsdale, AZ, USA, 2011.

[25] Francesco Lo Conti, Gabriele Minucci, and Naser Derakhshan. 2017. A Regularized Deep Learning Approach for Image De-Blurring. In IML '17: International Conference on Internet of Things and Machine Learning, October 17–18, Liverpool, United Kingdom. ACM, New York, NY, USA, pp. 1–55, 2017.

[26] R. Yan and L. Shao, "Blind Image Blur Estimation via Deep Learning," in IEEE Transactions on Image Processing, vol. 25, no. 4, pp. 1910–1921, April 2016.

[27] Anwesa Roy and et al., "Blur Classification and Deblurring of Images", International Research Journal of Engineering and Technology, Vol 04, Issue 04, Apr -2017.

[28] R. Shiva Shankar and et al., "A Novel Approach for Sharpening Blur Image using Convolutional Neural Networks", Journal of Critical Reviews, vol 7, Issue 7, 2020.

[29] Adaline suji and et al., "Classification of Malignant Melanoma and Benign Lung Cancer by using Deep Learning Based Neural Network", International Journal of Innovative Technology and Exploring Engineering, vol 9, issue 3, 2020.

[30] Yang H, Su X, Chen S, Zhu W, Ju C, "Efficient learning-based blur removal method based on sparse optimization for image restoration" PLoS ONE 15(3): 0230619, 2020.

[31] Sonu Jain and et al., "Image Deblurring from Blurred Images", International Journal of Advanced Research in Computer Science & Technology, Vol. 2, Issue 3, 2014.

[32] R. Aruna and et al., "An Enhancement on Convolutional Artificial Intelligent based Diagnosis for Skin Disease using Nanotechnology sensors", Journal of Nanomaterials, vol. 2022, pp. 1–6, July 2022.

[33] K. Srihari and et al., "Analysis of convolutional recurrent neural network classifier for COVID-19 symptoms over computerised tomography images", International Journal of Computer Applications in Technology, Inderscience Publishers, vol 66, issue 3–4, pp. 427–432, Jan 2022.

[34] K. Srihari and et al., "Ubiquitous Vehicular Ad-Hoc Network Computing using Deep Neural Network with IoT-Based Bat Agents for Traffic Management", Electronics, MDPI, vol 10, issue 7, pp. 785, March 2021.

[35] T. Kanagaraj and et al., "An innovative approach on health care using Machine learning", Intelligent System Algorithms and Applications in Science and Technology, Research Notes on Computing and

Communication Sciences, CRC Press, Taylor & Francis group, pp. 83–99, 2021.

[36] P. Poongodi and et al., "An Innovative Machine Learning Approach to Diagnose Cancer at Early Stage", Data Analytics in Bio Informatics- A Machine Learning Approach, John Wiley & Sons, pp. 313–337, 2021.

11

Method for Characterization of Brain Activity using Brain–Computer Interface Devices

Wilver Auccahuasi[1], Oscar Linares[2], Karin Rojas[3], Edward Flores[4], Nicanor Benítes[5], Aly Auccahuasi[6], Milner Liendo[7], Julio Garcia-Rodriguez[8], Madelaine Bernardo[9], Morayma Campos-Sobrino[10], Alonso Junco-Quijandria[10] and Ana Gonzales-Flores[10]

[1]Universidad Científica del Sur, Perú
[2]Universidad Continental, Perú
[3]Universidad Tecnológica del Perú, Perú
[4]Universidad Nacional Federico Villarreal, Perú
[5]Universidad Nacional Mayor de San Marcos, Perú
[6]Universidad de Ingeniería y Tecnología, Perú
[7]Escuela de Posgrado Newman, Perú
[8]Universidad Privada Peruano Alemana, Perú
[9]Universidad Cesar Vallejo, Perú
[10]Universidad Autónoma de Ica, Perú

Email: wauccahuasi@cientifica.edu.pe; olinares@continental.edu.pe;
krojas@utp.edu.pe; eflores@unfv.edu.pe; nbenites@unmsm.edu.pe;
aly.auccahuasi@utec.edu.pe; milnerdavid.liendo@epneuman.edu.pe;
julio.garcia@upal.edu.pe; mbernardo@ucv.edu.pe;
mcampos@autonomadeica.edu.pe; ajunco@autonomadeica.edu.pe;
agonzales@autonomadeica.edu.pe

Abstract

In recent years and with the technological evolution, we can perform tasks that years ago were impossible to develop. This is the case of being able to have records of brain activity, where expensive medical equipment is required, as

257

well as the use and exploitation of the signals were limited, today we can find solutions based on the use of brain–computer interface, based on low-cost devices, where you can make records of brain activity of various types of signals, these signals depend on the equipment, Another factor is the configuration of the electrodes that is required to make a good record, much of the equipment uses electrodes that are placed in the cerebral cortex, in this work we present the exploitation of a low-cost device, which uses a solid-state electrode placed on the forehead and with an armor that is placed on the head, which is easy to use, the equipment has a wireless connection, which allows transmitting the signal to different devices. As a result, we present the protocol of use of the equipment to have a good record, as well as to be able to prepare the data for later analysis; we present the different applications where brain activities can be recorded. The method is applicable, scalable, and adaptable to the different needs required.

11.1 Introduction

Technology allows us to evaluate new ways of analyzing the behavior of the brain, thanks to the electroencephalography equipment, nowadays commonly called BCI devices. Thanks to these devices, we can evaluate how the brain reacts to the activities that people perform. In the literature review, we found several works where they make use of these devices. We can indicate works where the behavior of the brain is evaluated in situations related to online classes in children, and evaluations are made mainly in the evaluation of the levels of attention and meditation. These studies are related to measuring how children react to the various situations of online classes [1]. We also found works related to the simultaneous use of virtual reality devices. The intention of the work is to evaluate how the brain reacts to rehabilitation situations, in a normal situation patients are viewing the images in a procedure called motor imagery to relearn how is the movement of their limbs. These images are presented consecutively so that the patient can identify the limbs if they correspond to the right or left side. The use of BCI devices is related to analyzing if the patient concentrates better when the images are projected with virtual reality glasses [2].

Continuing the use of BCI devices, in the analysis of brain behavior, we find works related to the evaluation of concentration and meditation levels in the realization of video games, these studies try to demonstrate the behavior of the brain when children play and measure mainly the levels of attention, so it is known how the brain reacts to these activities [3]. These same techniques related to the evaluation of the levels of attention and meditation, we found works related to rehabilitation exercises, mainly in the processes of performing exercises on stationary bicycles, to know the behavior of the brain when people are doing their exercises [4] [5].

The present investigation will analyze the use of prosthesis and its problems with the intervention of BCI in people with physical limitations it requires a method of invasive recording, so it is part of patterns of unrelated brain activities, so it has been evaluated how to control the grip and release of an upper limb prosthetic terminal device by classifying the electroencephalogram (EEG) data of the actual movement of grip and release of the hand, to perform the tests have been used five healthy people which has been recorded by Emotiv EPOC headset noninvasive consumer-grade, for this process was given orders isometric movements to participants which were compared by EEG data, we relate them to the movements performed to perform a recording of EMG used alternatively to then compare the data, within EEG has been increased signal/noise, special patterns, spectrally weighted (spec-CSP) for feature extraction, which allowed a good classification of EEG improving it up to 73.2% between grasp and release because the presented approach analyzes EEG signals of hand grasp and release movements with BCI intervention considered low-cost and noninvasive [6].

The present research work analyzed about asynchronous BCI that has been based on imagined speech considered as a tool for the control of external devices, issuing a message according to the user's need with a decoding of the EEG on imagined speech in order to correctly define the types of BCI should detect a continuous signal that is formed when you start to imagine words, for which it has been used five methods for feature extraction, by implementing asynchronous BCI on imagined speech, for which three datasets containing four different classifiers have been used. The results obtained are quite promising about the automation system for segmentation of the imagined words [7].

In the present research work we will analyze about the selection of the channels considered as relevant in order to produce an optimal subset of the EEG features considered as paramount for the reduction of computational complexity, reduction of overfitting, elimination of inconvenience on the users during the application, These signals have been taken from the scalp with extraction techniques and classification of the features considered as important in BCI systems based on motor imagery which are considered as high density with 64 channels for recording EEG signals, however when we applied it was verified that fewer channels have been required in the system configuration, Therefore, it is sought to determine the optimal number of channels for the use of BCI systems, for which the channel selection strategy has been proposed using a neuroevolutionary algorithm based on the optimization of the swarm of particles that were modified (MPSO), also using the use of feature extraction with spatial patterns, After the presentation of the method and its validation, 64-channel EEG recordings of four transhumeral amputees have been performed, signals from five motor imaging tasks have been acquired and introduced into a neuroevolutionary framework based on the envelopes after extraction[8].

In the present research will be analyzed about the fatigue considered as a factor in car accidents which can be measured by squinting, yawning, ignition control, and dispute, but so far the methodologies only evaluate about physical activities, so we propose a methodology where we separate the mental activities evaluated by EEG signals that will depend on the development of BCI, where it is considered essential to separate the signals from the brain, which has a large number of interconnected neurons, the model presented will change human contemplations, in this game plan will outline a transcendental electric brain banner, that is to say the evaluation was made with the person sitting with open eyes, changing the condition to normal for which a mental wave sensor was used with which brain indications are accumulated which were based on EEG with several repetitions where the signs of the groups have been changed sending the information via Bluetooth to the measurement divider section to verify the level of thinking [9].

In the present investigation, we will analyze about BCI applied in the motor images (MI) translated from EEG signals during the fulfillment of different tasks, alternative systems use high quality algorithms with which are trained with offline documents which is considered slow, they have a low quality heuristics but fast in real-time training but has a large classification error, for which we propose a new processing pipeline with a parallel learning and in real time on EEG signals along with a high quality algorithm but inefficient, which is realized by BCI and kernels, I consider techniques originated from computational geometry with transmission data handling by data summaries, such algorithm has adapted kernels in EEG handling by performing real-time continuous computation about common spatial pattern (CSP) feature extraction in a kernel representation of signal also improves the efficiency of CSP algorithm which is applied to the central kernel, then we have added in real-time data tests to the central kernel in order to achieve simplification we have focused on the CSP algorithm considered as classical [10].

The present research has evaluated about EEG signals which can reduce the performance of EEG analysis in the later stage about decision-making classified as BCI, it has been possible to appreciate about the problematic detection and elimination of artifacts for single channel BCI reading after this has been presented a mapping about the probability of EEG can be read in four different statistical measures such as entropy, kurtosis, asymmetry, and periodic waveform, so it is proposed an artifact removal that was based on standing waves used in probability threshold that are provided by the user, performing synthetic tests as real where the method showed superiority in artifact suppression with minimal distortion after the tests have been compared the results of both methods being a superior performance the proposed method [11].

In the present research, we analyze about BCI in the practice of existing communication between human and computer performing a self-regulation of the electrical activities of the brain, it has been based on EEG signals considering a form of application where you learn to control brain waves of μ-rhythm type (8–14 Hz) or β-rhythm (14–30 Hz) or potential related to P300 events or slow cortical potentials (SCP), extracting the signals using the processing technique, so the autoregulation of SCPs was analyzed using wavelet packets (WPA), which delivers an arbitrary time-frequency resolution, being able to analyze signals of stationary or nonstationary nature, with a better Fourier analysis time and a higher resolution with high frequency than the wavelet analysis, with the subimages of WPA where they have been analyzed with the logarithmic energy entropy, where the feature vectors are introduced a (MLP), for which the datasets Ia and Ib of BCI Competition 2003 have been used with which were tested then the comparison of MLP with k-NN and SVM has been performed where an accuracy of 92.8% and 63.3% was obtained in Ia and Ib data [12, 13].

In this work, we demonstrate a method to characterize the behavior of the brain by analyzing the electroencephalography signals captured by the BCI device. The signals [14] captured are delta, theta, low-alpha, high-alpha, low-beta, high-beta, low-gamma, and mid-gamma, which can be applied to the analysis of many applications in different areas.

11.2 Materials and Methods

The proposed method is characterized by describing the necessary steps to characterize the behavior of the electroencephalography signals, for which it is necessary to know how the brain behaves, how its behavior can be evidenced, what equipment can be used to record its activity, and what uses can be made using the data captured by the equipment.

Figure 11.1 describes the steps necessary to describe the method, from the description of the requirements to the use and applications of the electroencephalography signals.

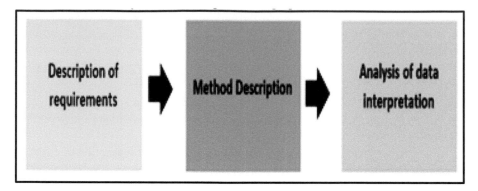

Figure 11.1   Description of the proposed method.

11.2.1 Description of the requirements

The requirements are characterized by the need to know the behavior of the human brain when performing various activities, to be able to analyze how the brain reacts to certain external influences. The use of equipment and electroencephalography signals is commonly used in the analysis of pathologies related to human behavior, In order to investigate new ways to evaluate the behavior, new equipment dedicated to record EEG signals in a practical way, these devices are commonly called BCI devices, called brain–computer interface, these devices perform the recording, analysis, and interpretation of EEG signals of people who use the device. In this paper, we describe the use of one of the low-cost BCI devices, model NeuroSKY MindWave, which has the property of being able to perform the recording through a single solid electrode. The device analyses the levels of attention and meditation at the time of recording the signals, as well as delta, theta, low-alpha, high-alpha, low-beta, high-beta, low-gamma, and mind-gamma signals. In the content, we will find the mechanisms to be able to configure the device to be able to perform the clean recording of EEG signals.

11.2.2 Method description

The description of the method is described below and is related to the configuration of the device, the mode of use from the initial configuration, and the applications where the various benefits of the NeuroSKY device can be exploited.

Figure 11.2 shows the NeuroSky model BCI device. Its main feature is that it is easy to use and practical; it uses an AAA battery power supply,

Figure 11.2 NeuroSKY BCI device.

which shows its low power consumption, another feature that makes it easy to use, is its physical structure is in the form of a headband, which is practical to place on people, in Figure 11.2 shows an image of the device.

Figure 11.3 shows how to use the BCI device, which is placed on the front of the person, indicating the sensor on the forehead, this form of physical configuration, allows to record through the device the behavior of the brain through various actions, the convenience of the device allows the records, in the figure shows the record when a person is in online classes, where the study intends to record the levels of attention and meditation, for which the device made the record of 4 h of class, which allowed us to make a continuous record.

Having configured the mode of use in people or patients, the next procedure is to perform the connectivity between the device and the computer, for which we indicate the following procedures:

Figure 11.4 shows the home screen of the device configuration. One of the characteristics of the BCI device is the connectivity that uses Bluetoot communication protocols, in order to configure the transmission mechanism, first the device and the computer or mobile equipment must be paired, having configured the means of signal transmission, the next step is to verify the

Figure 11.3   How to use the BCI device.

Figure 11.4   Welcome screen of the BCI device interface.

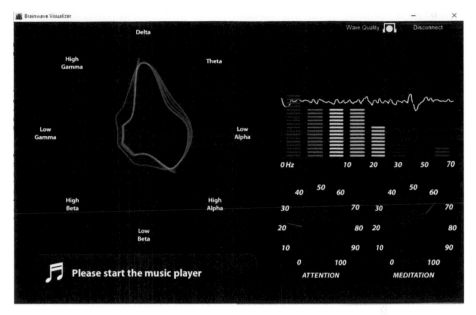

Figure 11.5 Application for device connectivity verification.

communication. This procedure is done through the use of an application where we demonstrate the connectivity, the level of signal that you have with the equipment and the integrity of the signals that are being received.

In order to verify that the device's information can be securely recorded, an application is used to corroborate that the information is received securely. Below is the image of the application that is recommended to be used to verify the configuration parameters, such as the signal level, which is represented as the signal that identifies the quality of the connection. In the upper right, a green image that identifies the good connection is verified; one of the qualities of the Bluetooth communication protocol is the distance between the two devices. To ensure the integrity of the data is required that the distance between the person who has placed the device on his head should be as close to the computer or the application, application, as shown in Figure 11.5.

Having configured the transmission mechanism of the signals, the next step is to verify if the signals are being received normally, understanding normality to be able to record the values recorded by the BCI device. In our case, it records attention and meditation values, followed by eight main EEG signals, such as delta, theta, low-alpha, high-alpha, low-beta, high-beta, and low mind-gamma: delta, theta, low-alpha, low-alpha, high-alpha, low-beta, high-beta, low-gamma, mind-gamma. Figure 11.5 shows an application that visually evidences the registration of the indicated signals, and we can observe that for the values of the levels of attention and meditation, the application presents a register in values from 0 to 100, for the values of the EEG signals, the application presents a visual diagram that identifies the values of the eight signals simultaneously, with this application it can be demonstrated that the signals are registered and visualized without problems.

The indicated procedure is performed sequentially, if at any time there is any unconventional situation, the recommended procedure is to perform the pairing between the BCI device with the computer, this procedure can be performed manually, with the assistant of the operating system, these manual configurations may take some time to connect, for which it is recommended to restart the device to improve communication, in this case, it is required to perform the conformation with the use of the application, in Figure 11.6 shows the flowchart of the procedure, where are the processes of device configuration, signal recording and signal analysis, in this process it is required to have the conformity of the signal recording, in this procedure the application indicated above is used, if the signals are in accordance we proceed with the use and exploitation of the signals, otherwise the device is reconnected, this operation is performed until it is considered that the signal recording is in accordance.

One of the characteristics of the BCI devices is the type of signal that is recorded, so the importance of protocol is of utmost importance. In this particular case, EEG signals are very small; in this sense, the device has the signal conditioning and amplification unit, one of the most serious effects is the interference of the medium; in order to eliminate these types of artifacts, it is necessary to comply with the procedures considered from the use by the person or patient, to the connectivity with the computer, it is important to indicate that one of the important tasks is the pairing through the Bluetoot protocol.

Figure 11.6 shows the flow diagram of the method for ensuring the integrity of the information from the recording device to the transmission of the signals.

One of the important characteristics of the methodology is described in Figure 11.7, where the architecture of the method is shown in order to have an idea of how the different agents involved in the use of the device are integrated; first, we have the requirements together with the patient, where we define the requirements for the choice of the best device to solve the problem and the requirements identified, then we select the best BCI device, where we have to identify the mode of use and the interaction with the patient. We perform the process of recording and collecting signals, having these signals

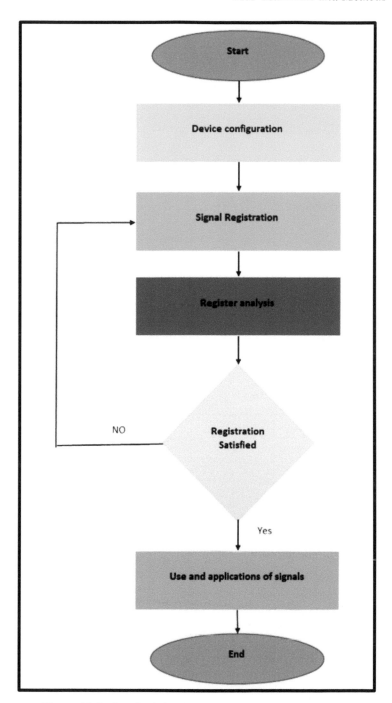

Figure 11.6 Received signal verification procedure flow chart.

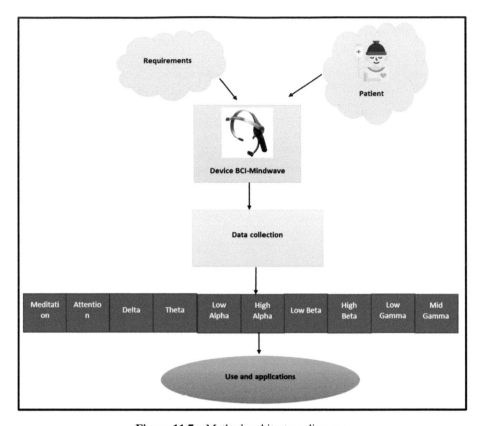

Figure 11.7 Method architecture diagram.

in different formats depending on the device and the configuration, then we have the procedure of separating the signals. Since the file provided by the devices are in CSV, TXT, DAT, and EXCEL, among others, we must have a mechanism for separating the signals for subsequent analysis depending on the applications that need to be used, as shown in Figure 11.7.

11.2.3 Analysis of data interpretation

Continuing with the explanation of the methodology, at this point, we have configured the device with the computer, where the storage of the information will be performed, resulting in the file with the recording of the signals. The following procedure is related to separating the general file into separate files, each identified with the signal. In this case, we have ten signals in the general file: two with the levels of attention and meditation and eight with

```
Time instant,
       1.  "Meditation",
       2.  "Attention",
       3.  "delta (0.5 - 2.75Hz)",
       4.  "theta (3.5 - 6.75Hz)",
       5.  "low-alpha (7.5 - 9.25Hz)",
       6.  "high-alpha (10 - 11.75Hz)",
       7.  "low-beta (13 - 16.75Hz)",
       8.  "high-beta (18 - 29.75Hz)",
       9.  "low-gamma (31 - 39.75Hz)",
       10. "mid-gamma (41 - 49.75Hz)"
```

Figure 11.8   Electroencephalographic signals recorded by the device.

characteristic signals of the EEG signals. Figure 11.8 shows the organization of the types of signals in the file with the general record.

To be able to perform the separation process, it is necessary to know the order in which the signals are represented in the general log file. In this case, in Figure 11.8, we show how the log files are organized. In Figure 11.9, we show how these sales are displayed in an Excel file and show how these signals are visualized in an Excel file, the original format of the file is CSV, the following procedure is related to the separation of the signals in separate files, for which we have several options for these processes, such as those related to the use of a programming language that can work with matrices to perform the separation.

Figure 11.9 shows the original data in CSV format, using the Excel program, in order to analyze and perform the procedure of separating the signals into independent vectors.

11.3 Results

The results that we present are related to the presentation of the signals in graphic form. For a complete recording of the signals, it is important to indicate that the proposed method is dedicated to presenting a mechanism for using BCI devices to record brain signals. The method presented and explained resorts to the use of the BCI device of the NeuroSKY model, which

	A	B	C	D	E	F
1	Time instant,"Meditation","Attention","delta (0.5 - 2.75Hz)","theta (3.5 - 6.75Hz
2	2021-06-13 18:19:21,0,0,"0,36","0,31","0,01","0,21","0,10","0,00","0,02","0,00"
3	2021-06-13 18:19:26,44,41,"0,16","0,32","0,14","0,12","0,06","0,08","0,09","0,04"
4	2021-06-13 18:19:32,38,100,"0,48","0,10","0,05","0,02","0,00","0,12","0,18","0,05"
5	2021-06-13 18:19:37,29,34,"0,14","0,53","0,10","0,02","0,07","0,08","0,05","0,02"
6	2021-06-13 18:19:42,38,40,"0,22","0,51","0,12","0,03","0,05","0,06","0,01","0,01"
7	2021-06-13 18:19:47,66,74,"0,97","0,01","0,00","0,00","0,01","0,00","0,00","0,00"
8	2021-06-13 18:19:52,29,53,"0,87","0,06","0,02","0,01","0,01","0,01","0,01","0,02"
9	2021-06-13 18:19:57,43,56,"0,49","0,26","0,02","0,01","0,04","0,08","0,09","0,00"
10	2021-06-13 18:20:02,27,57,"0,34","0,61","0,02","0,02","0,00","0,00","0,01","0,00"
11	2021-06-13 18:20:07,50,48,"0,20","0,31","0,19","0,02","0,04","0,02","0,18","0,04"
12	2021-06-13 18:20:12,100,26,"0,71","0,17","0,00","0,05","0,03","0,04","0,00","0,01"
13	2021-06-13 18:20:17,67,63,"0,87","0,05","0,01","0,02","0,00","0,01","0,03","0,01"
14	2021-06-13 18:20:22,74,61,"0,50","0,27","0,07","0,06","0,01","0,06","0,04","0,00"
15	2021-06-13 18:20:27,40,78,"0,34","0,04","0,15","0,20","0,11","0,12","0,03","0,00"
16	2021-06-13 18:20:32,24,43,"0,60","0,37","0,00","0,02","0,00","0,00","0,00","0,01"
17	2021-06-13 18:20:37,91,67,"0,56","0,11","0,11","0,03","0,04","0,10","0,04","0,02"
18	2021-06-13 18:20:42,63,38,"0,58","0,34","0,03","0,01","0,01","0,03","0,01","0,01"
19	2021-06-13 18:20:47,56,53,"0,13","0,69","0,01","0,05","0,03","0,05","0,04","0,01"
20	2021-06-13 18:20:52,56,51,"0,58","0,15","0,04","0,08","0,01","0,05","0,06","0,03"
21	2021-06-13 18:20:57,80,64,"0,84","0,01","0,04","0,02","0,00","0,02","0,05","0,02"
22	2021-06-13 18:21:02,78,67,"0,92","0,01","0,00","0,00","0,01","0,02","0,03","0,00"
23	2021-06-13 18:21:07,40,83,"0,29","0,32","0,05","0,08","0,01","0,09","0,02","0,14"
24	2021-06-13 18:21:12,61,54,"0,03","0,13","0,23","0,13","0,21","0,12","0,08","0,07"
25	2021-06-13 18:21:17,43,88,"0,56","0,27","0,04","0,08","0,01","0,03","0,00","0,00"
26	2021-06-13 18:21:22,63,57,"0,74","0,08","0,03","0,07","0,02","0,01","0,01","0,04"
27	2021-06-13 18:21:27,47,78,"0,90","0,03","0,01","0,04","0,00","0,01","0,01","0,00"
28	2021-06-13 18:21:32,67,74,"0,51","0,13","0,11","0,05","0,13","0,05","0,02","0,01"
29	2021-06-13 18:21:37,50,64,"0,92","0,01","0,02","0,00","0,01","0,01","0,01","0,01"
30	2021-06-13 18:21:42,4,90,"0,97","0,00","0,00","0,00","0,01","0,00","0,01","0,00"

Figure 11.9 CSV file format with the information from the signal recording.

uses a single solid electrode for recording, which presents a file with CVS extension. In this sense, we present the signals separately to demonstrate the method.

Having the general file, it is necessary to separate them, and we resort to the use of the computational tool MATLAB, which works with matrices. The process recommended by the tool is to load the file and then perform

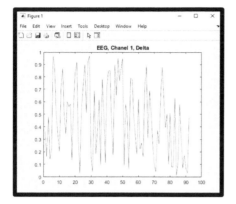
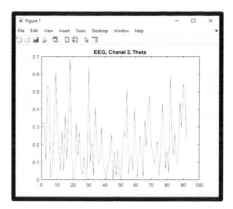

Figure 11.10 Delta signal recording and its graphical interpretation.

Figure 11.11 Theta signal recording and its graphical interpretation.

the separation into separate vectors for each signal. The original matrix is composed of ten vectors, divided into ten vectors separately and analyzed separately, and then we present the visualization of the signals separately.

When we separate the signals, the first vectors are those related to the levels of attention and meditation, so we will start analyzing from the vector found in position 3, which corresponds to the signal "delta." The signal is presented with its characteristic values and represented graphically with the plot function in MATLAB, as shown in Figures 11.10 – Figure 11.17.

The second vector is related to the "theta" signal in position 4 of the main matrix. Then we present the signals of the other vectors. In position 5 of the signal is the "Low-alph" signal. Position 6 is the signal that corresponds to the "high-alph" signal. In position 7, we have the signal that corresponds to the "Los-beta" signal. In position 8, we have the "high-beta" signal. In position 9, we have the "Low-gamma" signal; and finally, in position 9, we have the "low-gamma" signal. In position 7, we have the signal corresponding to the signal "los-beta," in position 8, we have the signal "High-beta," in position 9, we have the signal "low-gamma," and finally, in position 10, we have the signal corresponding to the signal "mind-gamma," then we present the signals corresponding to each of the identified signals, as shown in Figure 11.11.

The protocol presented, allows to obtain the signals that can be recorded depending on the BCI device that you may have. Although we cannot say that it is a medical device in general, we indicate that it is a device related mainly to researching the behavior of the brain. It is important to note that depending on the device, the signals to be acquired may vary, and we recommend analyzing in detail the characteristics of the devices to know the type of

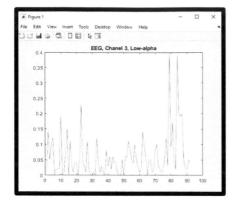

Figure 11.12 Low-alpha signal recording and its graphical interpretation.

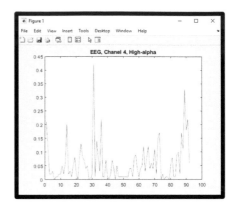

Figure 11.13 High-alpha signal recording and its graphical interpretation.

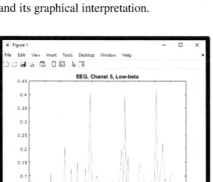

Figure 11.14 Low-beta signal recording and its graphical interpretation.

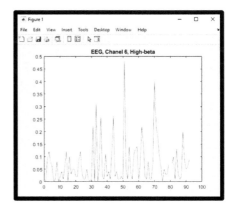

Figure 11.15 High-beta signal recording and its graphical interpretation.

communication, and the types of signals to acquire, among others, as shown in Figures 11.12 – Figure 11.17.

11.4 Conclusion

The conclusions we reached at the end of the explanation of the methodology are represented by three stages. The first is dedicated to the potential of the projects or research that can be done, all related to assessing the behavior of the brain, measured through its brain activity, which can characterize their behavior to a situation or action; we can indicate applications in education, where we try to measure the levels of behavior of the brain in different daily

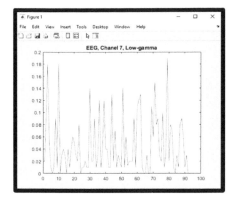

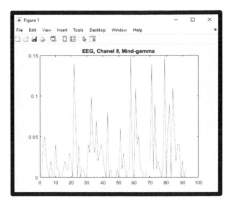

Figure 11.16 Low-gamma signal recording and its graphical interpretation.

Figure 11.17 Mind-gamma signal recording and its graphical interpretation.

situations related to study methods techniques, we can find which of the techniques applied, lead to a better understanding of the students and know how
it influences their level of attention, for these tasks it is necessary to know
if the students are attending the class, using the BCI device, we can register
these values, thus being able to evaluate which of the techniques used helps
the students more in their cognitive learning processes.

A second stage is considered to be able to solve a problem using BCI
devices, according to the benefits that the device can provide, these are
related to the type and mode of recording EEG signals, which is defined
from the different types of signals that can be recorded as well as the different channels that are counted, in a further analysis the more channels the
device has, the greater the number of signals to capture, which increases the
possible patterns to look for that can characterize the activity, it is important to note that in the market we find different BCI devices, Each of them
with particular characteristics, which may differ from each other, such as the
recording mode, the number of channels, the types of signals to record, the
type of power supply and the type of communication with other devices for
the transmission and recording of signals, we recommend analyzing the BCI
device, according to the needs and requirements needed to solve the problem
established, the choice of device will be by the level of complexity of the
problem to be solved, the cost required, and the potential in the analysis of
the signals that are the essence of BCI devices.

The third and last stage is characterized by the way the devices are used,
in the methodology we describe a procedure to ensure that the signal being
recorded is correct, we must take into account, by the nature of EEG signals,

which are very small values, it is necessary to have various electronic circuits that make possible the conditioning of the signal, as well as the application, one of the characteristics of these BCI devices; One of the characteristics of these BCI devices, which greatly help the use and applications of EEG signal exploitation, is the presentation of the signals in the best way because they come implemented with the indicated circuits, leaving the researchers the task of exploiting the signals, instead of being preoccupied with the circuit design. To help the device provide us with a better signal, we must ensure that the equipment is well configured, from the placement to the patient, as well as the connectivity with the computer, in order to have the signal as clean as possible, in relation to the noise that may be present at the input of the electrode. Another recommendation is the distance between the device and the computer, considering that the closer the equipment can be, the closer the communication protocol ensures the integrity of the signal, so that the signal received corresponds to the signal from the brain.

Finally, we indicate that the proposed method can be applied, through the use of the BCI device of the NeuroSKY model, for various applications, as well as it can be scaled, which will depend on the complexity of the problem to be solved, we must indicate that the choice of device depends on the problem to be solved, there is no single strategy to solve the problem, it will depend on the best analysis for the choice of the best components of the system to solve the problem.

Acknowledgments

We thank the authors for their collaboration in the research and thank each of them for their dedication, recommendation, and availability at the time of describing the procedures for the method.

References

[1] Herrera, L., Auccahuasi, W., Rojas, K., Hilario, F., Flores, E., Huaranja, M., & Sernaque, F. (2021). Study of mental states in online classes caused by COVID-19, using brain-computer interface. In CEUR Workshop Proceedings (Vol. 3010, pp. 30–35).

[2] Auccahuasi, W., Diaz, M., Sandivar, J., Flores, E., Sernaque, F., Bejar, M., ... & Moggiano, N. (2019, November). Design of a mechanism based on virtual reality to improve the ability of graduated motor imagery, using the brain computer interface. In Proceedings of the 5th International Conference on Communication and Information Processing (pp. 119–123).

[3] Auccahuasi, W., Díaz, M., Sernaque, F., Flores, E., Aiquipa, G., Rojas, G., ... & Moggiano, N. (2019, November). Analysis of the comparison of the levels of concentration and meditation in the realization of academic activities and activities related to videogames, based on brain computer interface. In Proceedings of the 5th International Conference on Communication and Information Processing (pp. 154–157).

[4] Auccahuasi, W., Flores, E., Sernaqué, F., Sandivar, J., Castro, P., Gutarra, F., & Moggiano, N. (2019, November). Technique for the comparison of concentration and meditation levels in the performance of rehabilitation exercises in bicycle, using virtual reality techniques and brain computer interface. In 2019 E-Health and Bioengineering Conference (EHB) (pp. 1–4). IEEE.

[5] Aiquipa, W. A., Bernardo, G., Bernardo, M., Flores, E., Sernaque, F., & Núñez, E. O. (2019, October). Evaluation of muscular functional recovery based on concentration and meditation levels through the use of the computer brain interface. In Proceedings of the 2019 2nd International Conference on Sensors, Signal and Image Processing (pp. 73–76).

[6] Lange, G., Low, C. Y., Johar, K., Hanapiah, F. A., & Kamaruzaman, F. (2016). Classification of Electroencephalogram Data from Hand Grasp and Release Movements for BCI Controlled Prosthesis. Procedia Technology, 26, 374–381. https://doi.org/10.1016/j.protcy.2016.08.048

[7] Hernández-Del-Toro, T., Reyes-García, C. A., & Villaseñor-Pineda, L. (2021). Toward asynchronous EEG-based BCI: Detecting imagined words segments in continuous EEG signals. Biomedical Signal Processing and Control, 65. https://doi.org/10.1016/j.bspc.2020.102351

[8] Idowu, O. P., Adelopo, O., Ilesanmi, A. E., Li, X., Samuel, O. W., Fang, P., & Li, G. (2021). Neuro-evolutionary approach for optimal selection of EEG channels in motor imagery based BCI application. Biomedical Signal Processing and Control, 68. https://doi.org/10.1016/j.bspc.2021.102621

[9] Kanna, R. K., & Vasuki, R. (2021). Advanced BCI applications for detection of drowsiness state using EEG waveforms. Materials Today: Proceedings. https://doi.org/10.1016/j.matpr.2021.01.784

[10] Netzer, E., Frid, A., & Feldman, D. (2020). Real-time EEG classification via coresets for BCI applications. Engineering Applications of Artificial Intelligence, 89. https://doi.org/10.1016/j.engappai.2019.103455

[11] Islam, M. K., Ghorbanzadeh, P., & Rastegarnia, A. (2021). Probability mapping based artifact detection and removal from single-channel EEG signals for brain–computer interface applications. Journal of Neuroscience Methods, 360. https://doi.org/10.1016/j.jneumeth.2021.109249

[12] Göksu, H. (2018). BCI oriented EEG analysis using log energy entropy of wavelet packets. Biomedical Signal Processing and Control, 44, 101–109. https://doi.org/10.1016/j.bspc.2018.04.002

[13] Dose, H., Møller, J. S., Iversen, H. K., & Puthusserypady, S. (2018). An end-to-end deep learning approach to MI-EEG signal classification for BCIs. Expert Systems with Applications, 114, 532–542. https://doi.org/10.1016/j.eswa.2018.08.031.

[14] Devi, V. A., & Naved, M. (2021). Dive in Deep Learning: Computer Vision, Natural Language Processing, and Signal Processing. In *Machine Learning in Signal Processing* (pp. 97–126). Chapman and Hall/CRC.

12

Streaming Highway Traffic Alerts using Twitter API

Jayanthi Ganapathy[1], Ramya Mohanakrishnan[1], Ayushmaan Das[1], and Fausto Pedro Garcia Marque[2]

[1]Sri Ramachandra Faculty of Higher Education and Research, India
[2]University of Castilla-La Mancha, Spain

Email: dr.gj2021@gmail.com; ramz28m@gmail.com;
dasayush5maan@gmail.com; FaustoPedro.Garcia@uclm.es

Abstract

Streaming analytics is one of the applications of big data. This technology has enhanced the digital world in many folds ranging from simple customer analytics in a spreadsheet to mobile app development for travel planning and route guidance. This work presents various traffic forecasting methods that involve real-time data analytics. The primary objective of this chapter is to present the analytics behind streaming real-time traffic data from Twitter. This study aims at building an application that would accept the city name from the user, then generate all the traffic-related tweets in the city within a period of seven days from the date of the search. This would then alert the user to avoid those particular routes where there is a blockage, accident, or any other obstacle. The tweets are fetched from Twitter API using TweePy. These tweets later need to be classified as "traffic" or "nontraffic" based on the model. The tweets also need to be preprocessed in order to improve the efficiency of the model. The tweets are also lemmatized in order to improve the scope of search and accuracy. The most essential part is the building of model which is preceded by vectorization. This involves the representation of tweets as an array of numbers for the machine to understand. The model is trained with a preclassified dataset and then is used to classify the tweets which were fetched earlier. This chapter presents the data processing pipeline for streaming analytics using Twitter API. Toward the end, the chapter summarizes the results and directions for future research.

12.1 Introduction

12.1.1 Twitter and Twitter API

The Twitter Inc, an American company, which was created by Jack Dorsey, Noah Glass, Biz Stone, and Evan Williams in March 2006. Twitter is a platform for users to share their thoughts and ideas publicly and privately. Twitter is a social networking service that enables users to interact with messages in the form of tweets. However, all these messages can widely serve as a big storehouse of information. Such type of data has been termed as big data. The tweets can be used to access information and day-to-day happenings about a particular place, individual, etc.

In this work, the tweets have been fetched, and a model was built to only obtain the tweets related to "traffic" in a city. On reading the tweets, the users will be able to understand the conditions of roads around them and, accordingly, avoid taking congested or blocked routes. API stands for Application Programming Interfaces. Twitter API allows a programmer to access Twitter in a unique way. One can use the functionalities without actually opening the application. This is widely used for information systems, machine learning purposes, etc.

12.1.2 Preprocessing

This involves the removal of unnecessary characters, symbols, or blocks of text from a file. Preprocessing of tweets is essential as it makes the further steps in classification and model building more efficient and accurate. Tweets usually contain emojis, hashtags, mentions, etc., which are not of any use. Thus, they are needed to be discarded.

12.1.3 Model building

A model is created using Python as the programming language. Many libraries, primarily scikit-learn, provide us with various libraries and tools for building a model and training it with some preclassified dataset. After appropriate training, prediction is carried out for a particular tweet, and they are classified as either "traffic" or "nontraffic." Various models are used for the purpose in order to cross-check their accuracy and see which one is giving

Current Date	From Date	To Date	No. of tweets collected
10/07/2022	1/07/2022	9/07/2022	984
22/02/2022	11/07/2022	22/07/2022	236
28/07/2022	23/07/2022	27/07/2022	1153

Figure 12.1   Tweets from Twitter API.

greater accuracy. The ROC curves for various models are plotted and interpreted. The greater the area of the curve, the greater the accuracy.

12.1.4 Web application

A user-friendly web application has been developed which enables users to stream and filter traffic-related tweets of the city of their choice. This has been achieved using StreamLit. StreamLit is an open-source framework that has been developed in Python Language. It is widely used to develop web apps for data science or machine learning projects. Fetching Tweets from Twitter API Twitter API is accessed using the aforementioned keys and tokens. After successful authentication, tweets are fetched using .search_tweets() method of TweePy. The queries like languages, removing retweets, city name, etc., are created and passed as a parameter to the method. Accordingly, tweets are fetched, which can be stored in a CSV or Excel file for future reference (Figure 12.1).

1.   Collecting tweets from Twitter API

2.   Preprocessing of tweets followed by lemmatization

3.   Vectorization and model building

4.   Design of data pipeline for preparation of tweets and machine learning workflow

5.   Deployment of the framework into a web-app

The tweets contained unwanted symbols, emojis, characters, or words that should be removed for efficient model building. This is known as preprocessing 13 tweets. This is followed by lemmatization. Stemming and lemmatization involve grouping together the inflected form of a particular word so that they can be analyzed as a single item. It is required to increase the efficiency of searching tweets and vectorization. The steps represent a piece of text as a combination of numbers for the machine to understand. In machine learning, vectorization is a stage in highlight extraction. The thought is to get a few distinct elements out of the text for the model to prepare by changing the text completely to mathematical vectors. Two types of vectorizers have been used for the purpose:

- Count vectorizer: Works on the bag of words model

- TF-IDF vectorizer: Better than count vectorizer as it considers the overall document weightage.

12.1.5 Model building, training, and classification

Scikit-learn (sklearn) is one of the most widely used Python library that provides all the basic tools required for data analysis. After vectorization, the model is trained using a preclassified dataset. The following approaches have been used:

- Logistic Regression, Naïve – Bayes Model, Random – Forrest Classifier After training, the model is now used to classify the earlier fetched tweets either as _traffic' or _nontraffic'. Various models have various accuracy scores. The ROC curves for various models are visualized using Matplotlib and Seaborn. The curve which has a greater area will have greater accuracy.

Creation of a web application The web application has been developed using StreamLit, one of the most commonly used packages for building applications related to machine learning purposes.
The application has multiple pages, namely:

- Home: default page that opens on running the application

- Custom Search: Users can search for traffic-related tweets from a particular city. Also, the user can choose the model, approach, and test-split size for complete versatility.

- Quick Search: for searching traffic tweets in some common cities on the click of a button.

- About: information about the app 5.8 Deployment of the App on a server StreamLit provides a free cloud service that enables us to deploy the develop app on its server using GitHub. The service has been termed as StreamLit cloud. On its deployment, the app can be used from anywhere through the link generated after deployment.

12.2 Related Works

The methods of forecasting traffic in the near future can be classified into major subdivision, such as

1. Deterministic model,

2. Nondeterministic,

3. Machine learning,

4. Deep learning, and

5. Unsupervised pattern mining techniques.

The primary factors considered in the selection of models are the dynamic and stochastic nature of the physical traffic flow. For example, modeling of traffic flow using parametric approaches is successful in different operational settings when traffic time series is a stationary process, while another broad category of methods that are successful even though traffic conditions remain nonstationary includes machine learning approaches using neural networks as reported in [1-4].

Deterministic models are well suited for traffic forecasting operations in shorter horizons under different settings, whereas nonlinear models are well suited for prediction in longer horizons between 30 and 60 minutes in the future. Thus deterministic models arc well suited as long the time series traffic flow is linear and stationary.

The time series traffic forecasting considering the physical traffic flow as a single variable becomes a univariate model while forecasting traffic considering more than one variable, such as flow rate, occupancy, speed, etc., becomes a multivariate traffic flow [2, 4, 6-9]. The selection of features from time series traffic flow becomes an interesting phenomenon in case of nondeterministic models. Neural nets are robust elemental features with intense computation mechanism that executes parallelism in the computation of raw facts [10-13]. They are highly featured with feature extraction, preprocessing, dimensionality reduction, etc. In recent research, spatial and temporal dynamics of physical traffic flow is at most focus. Deep learners are highly potential to incorporate spatial and temporal dependence at a minimal cost besides the difficulties experienced in different operational settings [9, 11-13]. Moreover, system stability and model portability are very high in case of automated deep learners [9]. The following section highlights the various factors influencing the selection of methods for short-term traffic forecasting

Table 12.1 Selection of models and methods based on factors affecting the physical traffic flow.

Physical dynamics of traffic flow	Traffic forecasting models and methods[#]				
	D	*ND*	*ML*	*DL*	*PM*
Operational traffic condition (stationary)	•	✓	✓	✓	✓
Random traffic condition	•	✓	✓	✓	✓
Traffic prediction (5–15 min)	✓	✓	✓	✓	✓
Traffic prediction (above 30 min)	✗	✗	✓	✓	✓
Flow rate as a univariate traffic variable	✓	✓	✓	✓	✓
Flow rate and speed as multivariate traffic variable	•	✗	✓	✓	✓
Dimensionality reduction and data loss	✓	✓	✓	•	•
Feature selection	•	•	✓	✓	✓
Feature extraction	•	•	•	✓	✓
Spatial and temporal dependency	✗	✗	✗	✓	✓
Model and system portability	•	•	•	✓	✓
Training and validation loss (error)	High	High	High	Optimal	Less
Scalability of forecasting model	•	✓	✓	✓	✓
Run time (computational complexity)	Dependent			Independent	

Factors agreed: ✗: not agreed always •: never agreed ✓: agreed always

[#]Deterministic (**D**); nondeterministic (**ND**); machine learning (**ML**); deep learning (**DL**); pattern mining (**PM**).

12.3 Background

In our recent work, 2017–2022 [5–9] travel time-based traffic information sequence was formulated and implemented in a traffic information sequence mining framework. The sequential pattern mining framework [14, 15] was developed for the prediction of traffic flow on highways using the data set recorded at the centralized toll center. Real-time traffic volume data for 52 weeks is collected at a centralized toll system comprising all toll collection centers at three different sites in Chennai city, namely, (i) Site 1: Perungudi-Seevaram, the entry toll plaza (ii) Site 2: ECR link road, and (iii) Site 3: Egattur, the exit toll plaza. The data services of these three sites are under the authority of Tamil Nadu Road Development Corporation (TNRDC) in Chennai, Tamil Nadu, India. The research findings reported that traffic volume on highways could be predicted by mining travel time-based traffic information sequences, and it is feasible to deploy the framework in any suitable location. Table 12.1 presents the various factors influencing the selection of traffic forecasting methods.

The availability of historical traffic flow rate and connectivity of sites has motivated us to formulate the following objectives.

1. To capture the dynamics of the physical traffic flow by an Extract-Transform-Load (ETL) data pipeline design for the representation of raw traffic count.

2. To design a machine learning pipeline that augments the traffic sequence mining framework with vehicle speed based on multi-criteria decision-making support for profiling the highway traffic.

3. Design an analytic pipeline to disseminate dynamic traffic information in successive time instances and operate the vehicular traffic with the help of an interactive dashboard.

12.4 Objectives

Collecting tweets from Twitter API

- Preprocessing of tweets followed by lemmatization.

- Vectorization and model building.

- Design of data pipeline for preparation of tweets and machine learning workflow.

- Deployment of the framework into a web-app.

The primary objective of this project is to build an application that would accept the city name from the user, then generate all the traffic-related tweets in the city within a period of seven days from the date of search. This would then alert the user to avoid those particular routes where there is a blockage, accident, or any other obstacle.

The tweets are fetched from Twitter API [16, 17] using TweePy. These tweets later need to be classified as "traffic" or "nontraffic" based on the model.

The tweets also need to be preprocessed in order to improve the efficiency of the model. The tweets are also lemmatized in order to improve the scope of search and accuracy.

The essential part is building the model, which is preceded by vectorization. This involves the representation of tweets as an array of numbers for the machine to understand. Next, the model is trained with a preclassified dataset, then is used to classify the tweets fetched earlier. The aim is to obtain maximum accuracy and hence, multiple methods are used.

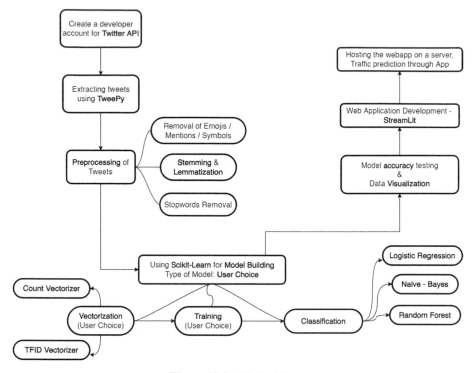

Figure 12.2   Methodology.

Ultimately, the development of web-app comes into action. StreamLit serves as one of the best open-source frameworks for this purpose. The aim was to make the app interactive as well as versatile, allowing the users to execute a search based on their choice.

The aim is to obtain maximum accuracy; hence, multiple methods are used. Ultimately, the development of web-app comes into action. StreamLit serves as one of the best open-source frameworks for this purpose. The aim was to make the app interactive as well as versatile, allowing the users to execute a search based on their choice.

12.5 Methodology

The sequential work methodology of the project involves fetching tweets followed by preprocessing, model building, and training – classification and data visualization. Ultimately, the concepts are integrated and deployed in the form of a web application using StreamLit (Figure 12.2).

Table 12.2   Fetching monthly tweets for testing.

Current date	From date	To date	No. of tweets collected
11/08/2022	1/08/2022	9/08/2022	884
22/08/2022	11/08/2022	22/08/2022	136
28/08/2022	23/08/2022	27/08/2022	1253

Developers account in Twitter

A developer's account must be created in order to access the Twitter API. On successful creation of the account, an app must be created within a project. The app must be given an authentication of version *OAuth1.0*, enabling us to stream tweets from Twitter API. The elevated level of access can be applied for, which is granted after a short period of time. The following features of the app must be stored safely for future reference:

- API key

- API key secret

- Bearer token

- Access token

- Access token secret

Fetching tweets from Twitter API

Twitter API is accessed using the aforementioned *keys* and *tokens*. After successful authentication, tweets are fetched, as shown in Table 12.2, using *.search_tweets()* method of *TweePy*. The queries like languages, removing retweets, city names, etc., are created and passed as a parameter to the method. Accordingly, tweets are fetched, which can be stored in a CSV or Excel file for future reference.

Text preprocessing and lemmatization

Text preprocessing is an essential task in processing tweets. The tweets may contain unwanted symbols, emojis, characters, or words that should be processed to remove for efficient model construction. This process may involve use of modern tools and techniques in Python. This is followed by *lemmatization*. Stemming involves grouping together the inflected form of a particular word, so that it can be analyzed as single item. It is required for increasing the efficiency of searching of tweets and vectorization.

Vectorization

Vectorization involves the representation of a piece of text as a combination of numbers for the machine to understand. In machine learning, vectorization is a stage in highlight extraction. The thought is to get a few distinct elements out of the text for the model to prepare by changing text completely to mathematical vectors.

Two types of vectorizers have been used for the purpose:

- *Count Vectorizer*: Works on the bag of words model

- *TF-IDF Vectorizer*: Better than count vectorizer as it considers the overall document weightage.

Model building, training, and classification

Scikit-Learn (sklearn) is one of the most widely used Python libraries that provides all the basic tools required for data analysis. After vectorization, the model is trained using a preclassified dataset. The following approaches have been used:

- Logistic regression

- Naïve-Bayes model

- Random – Forrest classifier

After training, the model is now used to classify the earlier fetched tweets either as "traffic" or "nontraffic." Various models have various accuracy scores. The *ROC curves* for various models are visualized using *Matplotlib* and *Seaborn*. The curve which has a greater area will have greater accuracy.

Creation of a web application

The web application has been developed using *StreamLit*, one of the most commonly used packages for building applications related to machine learning purposes. The application has multiple pages, namely:

- Home: default page that opens on running the application

- Custom search: user can search for traffic-related tweets from a particular city. Also, the user can choose the model, approach, and test-split size for complete versatility.

- Quick search: for searching traffic-tweets in some common cities with the click of a button

- About: information about the app

Deployment of the app on a server

StreamLit provides a free cloud service that enables us to deploy the developed app on its server using GitHub. The service has been termed as StreamLit cloud. On its deployment, the app can be used from anywhere through the link generated after deployment.

12.6 Technologies Used

Python

Python is a high-level and one of the most popular programming languages, first designed by Guido van Rossum which was first released in 1991. It has a widescale application and can be used for various tasks like data analysis, machine learning, building websites, automating tasks, etc.

Python is considered as one of the best languages for machine learning purposes because of the following:

- simple and consistent

- flexible as well as platform independent

- has a number of Libraries for machine learning purposes

- wide community of developers

TweePy

TweePy is a Python package that is open-source and is used by programmers to access the Twitter API. It comprises various types of classes and methods representing Twitter's models and API endpoints. It can give us almost all kinds of functionalities provided by Twitter API.

Before fetching tweets, one must use *ConfigParser* to read the keys and tokens of our Twitter API app stored in some file, generally "config.ini."

Pandas

Pandas is an open-source Python package. It is built on top of *NumPy*, which is another Python package. It is widely used for machine learning tasks. It serves the purpose of data cleansing, normalization, merges, joins, reading and writing data, etc. Various types of files can be read as dataframes, making it easy for data manipulation and analysis.

In this project, pandas have been used to read tweets from stored CSVs which are later used for displaying information and data visualization. In addition, some pandas functionalities also support limited data visualization.

Scikit-learn

Scikit-Learn is a library in Python used for machine learning and data analysis purposes. It provides all sorts of tools required for model building, including classification, regression, clustering, score matrices, vectorization, etc. It has been built on top of *NumPy*, *SciPy*, and *Matplotlib*. It is open-source and accessible to everybody.

In this project *logistic regression*, *random forest classifier,* and *Naïve-Bayes model* algorithms have been used from the library. Also for vectorization, scikit-learn's *Count Vectorizer* and *TF-IDF Vectorizer* have been implemented.

Natural language toolkit (NLTK)

NLTK is a Python toolkit used for natural language processing (NLP) [18] in Python. NLTK supports various functionalities like parsing, tokenization, stemming, lemmatization, etc. Lemmatization involves the conversion of the word into its base form. The extracted word is termed as *lemma*. The lemmas of words can be obtained using NLTK's *WordNetLemmatizer.*

Matplotlib and Pyplot

Matplotlib is used for plotting graphs and curves in Python. It is used in combination with *NumPy* and *Pandas* for data visualization in the form of graphs, plots, and charts.

Pyplot is a module within Matplotlib that provides MATLAB-like interface. Pyplot has various types of graphs like bar, pie, histograms, etc. Values can be obtained from any sort of datasets and graphs can accordingly be plotted.

Seaborn

Seaborn can be used as an alternative to Matplotlib. It uses Matplotlib underneath for data visualization. It provides some additional plots and a prettier output than the basic plots. Plots like Violin Plots, Swarm Plots, etc., can be used through seaborn.

StreamLit

StreamLit is a Python tool generally used to develop web applications specifically for machine learning and data visualization purposes. The app can be written in the same way and syntax in which we write Python code. StreamLit provides a wide range of features in the form of *widgets*. There are various types of inputs, graphs, and even markdowns that allow us to write a chunk of code in the form of HTML. Thus, it is versatile and supports a wide range of Python libraries including scikit-learn.

StreamLit Cloud is an app hosting cloud-based service provided by StreamLit for free. The app needs to be uploaded into a GitHub repository and then needs to be deployed into the cloud. Once deployed, it can be accessed through the link provided by the cloud.

12.7 TWITTER: Natural Language Processing

12.7.1 Vectorization

Word vectorization is a part of natural language processing (NLP) and is one of the mandatory steps in machine learning. This is required because our interpreters cannot understand words by itself. Vectorizers are used to map words or phrases in the vocabulary to a corresponding vector of real number.

Two different types of vectorizers have been used in this work: (i) CountVectorizer and (iii) TF-IDF vectorizer.

12.7.2 Terminologies

Corpus: collection of documents.

Vocabulary: collection of all the unique words within the corpus.

Document: every individual block of text (in this project and single tweet).

Word: every individual word within a document.

12.7.3 CountVectorizer

CountVectorizer is a part of the scikit-learn library of Python. It is based on the *bag of words* model. It can be used to convert a given word into a vector based on the frequency of each word in the entire block of text. It has a number of disadvantages, some of which are:

The words which are abundantly present in the vocabulary are simply considered as significant. Discrimination between more important words and less important words is lacking in this model. It is unable to study the similarity between words or phrases.

TF-IDF Vectorizer

TF-IDF stands for term frequency-inverse document frequency. TF is a measure of the frequency of a term in a document:

$$tf = \frac{n}{N} \quad tf = \frac{n}{\text{Number of terms in the document}} \tag{12.1}$$

	A	B	C
1	timestamp	text	class
2	May 5th 2018, 17:14:04.000	And some folks believe NYC got it right on VZ Any city using a police e	non_traffic
3	May 5th 2018, 16:25:23.000	When you find out last minute that the bus stop youre leaving NYC fr	non_traffic
4	May 5th 2018, 16:23:42.000	Any chance you would be open 30 mins later On our way up from NYC	non_traffic
5	May 5th 2018, 16:12:23.000	5BoroBikeTour 2018 is this SundayMay 6 A 40mile ride thru NYC s 5 b	traffic
6	May 5th 2018, 16:05:57.000	NYC is a traffic hellhole Chicago has better beer and baseball	non_traffic
7	May 5th 2018, 15:43:06.000	Something about NYC traffic that really makes me	non_traffic
8	May 5th 2018, 15:30:07.000	Heres this evenings train and traffic update Have a wonderful night N	non_traffic
9	May 5th 2018, 14:51:47.000	Marathon watching Marvel movies and all I can think about is who pa	non_traffic
10	May 5th 2018, 14:47:34.000	NewFad DTE uptown Embassy DOT Not sure what	non_traffic
11	May 5th 2018, 14:35:34.000	Touched down In NYC Traffic is crazy so I just hopped on the train fro	non_traffic
12	May 5th 2018, 14:11:40.000	Chris Hedges calls attention to the algorithms of Facebook Google an	non_traffic
13	May 5th 2018, 14:08:33.000	NYC Ferries are indeed nice but they should cost 6 at least thats what	non_traffic
14	May 5th 2018, 13:33:33.000	I wish the trains were more reliable in Nyc Not that Id start taking it a	non_traffic
15	May 5th 2018, 13:33:16.000	Planning on being in NYC this weekend Have fun and plan ahead with	non_traffic
16	May 5th 2018, 13:27:21.000	Senpai yeah like NYC was crazy but at least TM usually has their shit t	non_traffic
17	May 5th 2018, 13:23:13.000	The mayor can Add bus lanes Enforce bus lanes Implement transit sig	non_traffic
18	May 5th 2018, 13:18:41.000	Pedestrian intervention through traffic disruption in Greenwich NYC S	non_traffic
19	May 5th 2018, 13:03:15.000	Closed due to major event in Nyc on 2nd Ave SB between 14th St and	traffic
20	May 5th 2018, 12:54:51.000	DOT Id nominate the soon to have a protected bike lane and has very	non_traffic
21	May 5th 2018, 12:53:16.000	Ahh remember when you use to be able to drive somewhere in NYC o	non_traffic
22	May 5th 2018, 12:36:08.000	Traffic was slowed to a turtles crawl on the today Its actually a tortoi	non_traffic
23	May 5th 2018, 12:33:25.000	In NYC this weekend Plan ahead with our traffic advisory as there lots	non_traffic

Figure 12.3   The training dataset saved as CSV File.

Here, *n* is the number of times term *t* occurs in the document *d*. *N* is the total number of terms in the document. Every document has a separate TF value. Thus, this approach also provides the importance of a word in a document. TF-IDF is considered better than CountVectorizer because it successfully provides the significance of a word. Thus, nonimportant words can be removed for more efficient model building and lesser input size.

12.8 Results and Discussion

12.8.1 Exploratory analysis on datasets

The source used for training the Twitter data set is used from https://github.com/SivagurunathanV/Traffic-Detection-from-Twitter-using-Spark/blob/master/src/twitter_traffic_data_static.csv (Figure 12.3).

Step 1: The above training dataset was read using Pandas and the "class" column was converted into numerical column – 1 for traffic and 0 for nontraffic.

Step 2: The tweets were fetched from Twitter API using TweePy and stored in the computer memory in the form of a CSV file (Figure 12.4).

Step 3: The final dataset after the fetched tweets were, preprocessed, lemmatized, vectorized, and classified successfully by the model we designed (Figure 12.5).

	A	B	C	D	E	F
		username	tweet	location	date	time
	0	mattgrocoff	@DirtyTesLa @Tesla @elonmusk I just told my Tesla to	Ann Arbor, MI,	09-07-2022	01:07:49
	1	BoojiBoy6	@tentwentysixpm Living in NYC is expensive and finding a	Brooklyn, NY	09-07-2022	00:46:43
	2	CynfullySweetXO	@Slim_Luck_Alex drove thru NYC today ðŸ¤¬ I do like it u	Nowhere	08-07-2022	23:41:17
	3	jordonaut	@lantzarroyo Iâ€™m particularly scared for my folks in de	Seattle, WA	08-07-2022	23:02:33
	4	En_AmbientMusic	City Rain Traffic Sounds for Sleep and Study | ASMR		08-07-2022	22:49:18
	5	NYC81966570	@JohnnyGoodberry @alefeusch @wealth I think a lot of people's opinior		08-07-2022	22:48:19
	6	Dzollo_	@NYCMayorsOffice @nycgov @NYPDPC	New York, USA	08-07-2022	22:09:15
	7	ZalezVickie	@sanjeeva7 @casgroenigen05 @javroar Unless u in NYC. Them mfs smar		08-07-2022	22:03:36
	8	lolaxlachiva	Okay but nyc traffic ðŸ˜„		08-07-2022	21:50:39
	9	TVariunessKing	NYC needs trams. Replaces these buses that get stuck in t	40.7236448,-74	08-07-2022	21:50:09
	10	JMartinezNYC	@TransitNinja205 @Ollie_Cycles Weâ€™re not casting	New York, NY	08-07-2022	21:31:41
	11	Songbird99	@NYCTSubway @MTA @NYC_DOT There's a crazy white	Hooberbloob F	08-07-2022	21:23:56
	12	NYPDnews	If you are planning on spending time in NYC this	New York City	08-07-2022	21:00:09
	13	BIGKay95	NYC is exactly like the movies when it comes to this traffic	Tonawanda, N'	08-07-2022	20:43:55

Figure 12.4 CSV file: fetched tweets.

username	tweet	location	date	time	processed_tweet	predicted_class
mattgrocoff	@DirtyTesLa @Tesla @elonmusk I just told n	Ann Arbor,	2022-07-09	01:07:49	i just told my tesla to take my daugh	0
BooiiBov6	@tentwentysixpm Living in NYC is expensive	Brooklyn, N	2022-07-09	00:46:43	living in nyc is expensive and finding a	0
AquilesMp	One thing about NYC it's that not matter the	SomeWher	2022-07-03	09:36:37	one thing about nyc its that not matte	0
DutchKillsCivic	@NotifyNYC: .@FDNYAlerts Three Alarm Fir	Long Islanc	2022-07-03	08:28:43	three alarm fire th road amp nd stree	1
NotifyNYC	.@FDNYAlerts Three Alarm Fire: 85th Road 8	New York (	2022-07-03	07:40:18	three alarm fire th road amp nd stree	1
MrAFelix	@diemauerthewall Struck me since living hei	Lübben (Sp	2022-07-03	07:02:56	struck me since living here when i wa	0
TotalTrafficNYC	Accident. Two lanes blocked in #NYC:Onthel	New York (	2022-07-03	06:25:43	accident two lanes blocked in nycontl	1
cycling_nyc	@StreetwallNY @NYC_DOT @NYSDOT Encc	New York,	2022-07-03	06:08:31	encouraging washington heights resi	0
TotalTrafficNYC	Accident. Two lanes blocked in #NYC:OnThe	New York (	2022-07-03	05:40:43	accident two lanes blocked in nycontl	1
1Goodfriend12	@nypost Illegal Motorbikes on NYC Streets has reached	2022-07-03	04:53:18	illegal motorbikes on nyc streets has	0
drjamima	I'm thinking traffic stops are not going to be	Oregon	2022-07-03	04:12:43	im thinking traffic stops are not going	0
TotalTrafficNYC	Accident. Two lanes blocked in #NYC:OnThe	New York (	2022-07-03	03:45:43	accident two lanes blocked in nycontl	1
TotalTrafficNYC	Accident. Left lane blocked in #NYC:OnHenr	New York (	2022-07 03	03:25:43	accident left lane blocked in nyconhei	0
CovfefeAnon	@hjeutysd "Hey, NYC has murders and stabhings on the	2022 07-03	02:38:27	hey nyc has murders and stabbings or	0

Figure 12.5 CSV file: final result after preprocessing and classification.

Step 4: Data visualization

Graphs depicting the ratio of tweets classified as traffic to the total volume of tweets fetched (Figure 12.6).

The above graphs depict that although a large volume of tweets is fetched from Twitter API, all the tweets do not significantly ensure that they relay information about traffic conditions in the specified city (Figure 12.7).

12.8.2 ROC curve and AUROC

The ROC plots are receiver operating characteristic curves which depict the performance of the classification models and various levels of threshold. They use two essential parameters:

- True positive rate (TPR)

- False positive rate (FPR)

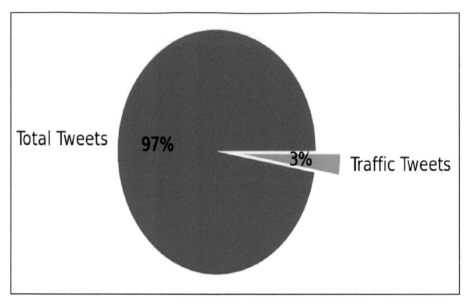

Figure 12.6   Pie chart.

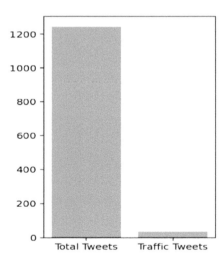

Figure 12.7   Bar plot.

$$\text{True positive rate } = \frac{\text{tp}}{\text{tp} + \text{fn}} \tag{12.2}$$

$$\text{False positive rate } = \frac{\text{fp}}{\text{fp} + \text{tn}} \tag{12.3}$$

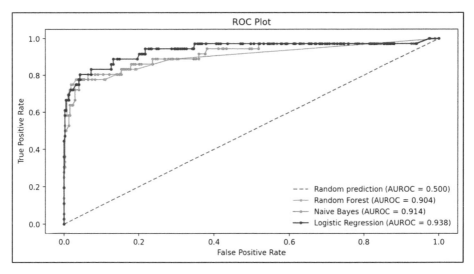

Figure 12.8 ROC plot.

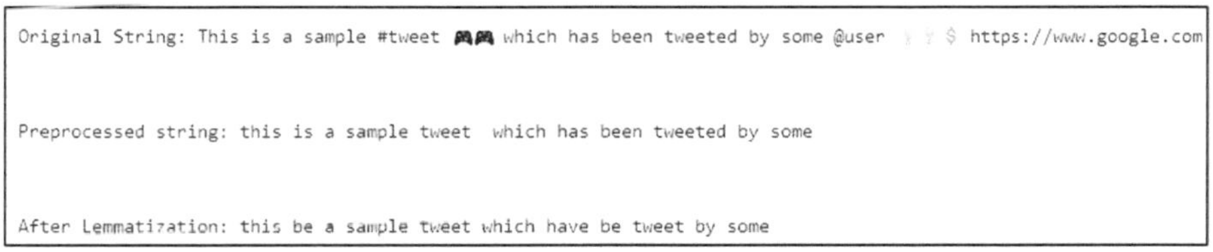

```
Original String: This is a sample #tweet 😀😀 which has been tweeted by some @user  ↑ ↗ $ https://www.google.com


Preprocessed string: this is a sample tweet  which has been tweeted by some


After Lemmatization: this be a sample tweet which have be tweet by some
```

Figure 12.9 Sample output: functionality of preprocessing tool.

"tp" denotes true positives; "fp" denotes false positives; "tn" denotes true negatives; "fn" denotes false negatives.

AUROC stands for area under a ROC curve. The greater the area under the curve, the greater our model's accuracy. Figure 12.8 shows the ROC plots

12.8.3 Preprocessing outputs

The preprocessing involves the removal of items such as symbols like hashtags, emojis, mentions, links, etc. Also, the output was later lemmatized as shown in (Figure 12.9); the words got converted to their lemma. Sample output showing the difference between fetched tweets before and after preprocessing (Figure 12.10).

tweet	processed_tweet
@DirtyTesLa @Tesla @elonmusk I just told my Tesla to take my daughter and me from supercharger in New Rochelle to Upper West Side of Manahattan. It went through towns, construction, traffic and down Hudson Pkwy with zero disengagements!! All while NYC drivers tried best to kill us. #FSDBeta https://t.co/IXK4mJrgrh	i just told my tesla to take my daughter and me from supercharger in new rochelle to upper west side of manahattan it went through towns construction traffic and down hudson pkwy with zero disengagements all while nyc drivers tried best to kill us fsdbeta
Living in NYC is expensive and finding an apartment is a mess but I don't find it stressful. What I do find stressful is having to sit in traffic and driver everywhere. I like visiting LA for the food/culture but would never want to live D5 @Slim_Luck_Alex drove thru NYC today 😫 I do like it upstate but man oh man, traffic is such a nightmare	living in nyc is expensive and finding an apartment is a mess but i dont find it stressful what i do find stressful is having to sit in traffic and driver everywhere i like visiting la for the foodculture but would never want to drove thru nyc today i do like it upstate but man oh man traffic is such a nightmare
@lantzarroyo I'm particularly scared for my folks in densely populated areas with poor ventilation (cities like NYC) that have high traffic in and out of there. Our public health system isn't doing as much as it COULD be. City Rain Traffic Sounds for Sleep and Study | ASMR Ambience | Relaxing ... https://t.co/hEj177VwgU via @YouTube City Rain Traffic Sounds for Sleep and Study #rainambience #rainvideos #rain #city #citytraffic #cityambience #loveny #ilovenyc #nyc	im particularly scared for my folks in densely populated areas with poor ventilation cities like nyc that have high traffic in and out of there our public city rain traffic sounds for sleep and study asmr ambience relaxing via city rain traffic sounds for sleep and studyrainambience rainvideos rain city citytraffic cityambience loveny ilovenyc nyc

Figure 12.10   Difference between the tweets before and after preprocessing.

Figure 12.11   Sample input.

12.8.4 Test inputs and outputs

"Los Angeles," a city in the United States was inputted in the text box. The split size was set to 0.3 and TF-IDF – Naïve-Bayes combination was chosen for classification (Figure 12.11).

The successfully fetched tweets from the API were displayed in the form of a table which can later be downloaded as a CSV file for further reference (Figure 12.2).

	username	tweet	date	time
0	TotalTrafficLA	Brush fire has ONramp blocked in #SouthLA on 110 (I-110 Hbr Fwy) NB at Gage Ave, stopped traffic bac	2022-07-26	78643000000
1	TotalTrafficLA	Accident. Shoulder in #Torrance on 405 NB at Normandie Ave, stopped traffic back to Carson St #LAtra	2022-07-26	57343000000
2	TotalTrafficLA	Crash carpool lane in #Riverside on I-215 NB after Center St (Exit 36), stopped traffic back to Hwy 60/Hv	2022-07-25	83743000000
3	TotalTrafficLA	Brush fire two right lanes blocked in #Pacoima on I-5 SB at Hwy 118/ Paxton St, stopped traffic back to	2022-07-25	75044000000
4	TotalTrafficLA	Accident cleared in #RanchoCucamonga on I-15 NB at Summit Ave, stopped traffic back to I-210 #LAtra	2022-07-25	6945000000
5	TotalTrafficLA	Accident cleared in #SouthLA on 110 (I-110 Hbr Fwy) SB at 51st St, stopped traffic back to Adams Blvd #	2022-07-24	22543000000
6	TotalTrafficLA	Accident. Two middle lanes blocked. in #SouthLA on 110 (I-110 Hbr Fwy) SB at 51st St, stopped traffic t	2022-07-24	21643000000
7	TotalTrafficLA	Accident cleared in #Topanga on Pacific Coast Hwy (N) SB at Coastline Drive, stopped traffic back to La	2022-07-24	14443000000
8	TotalTrafficLA	Accident. Shoulder in #Arleta on I-5 SB at Osborne St, stopped traffic back to Hwy 118/ Paxton St #LAtr	2022-07-22	52243000000

Figure 12.12   Fetched tweets.

SHOW ACCURACY SCORES FOR VARIOUS MODELS

	Logistic Regression	Naive-Bayes	Random Forest
Test Score	0.9510	0.9534	0.9608
Train Score	0.9600	0.9737	0.9968
Accuracy Score	0.9510	0.9534	0.9608

Figure 12.13   Accuracy scores.

Trafic tweets fetched	Tweets classified as 'Traffic'	Download the Tweets as a CSV
1056	48	

Figure 12.14   Metrics.

Different algorithms are seen to have different accuracy scores, as shown in Figure 12.13. This is because each one has a unique way of working. The one with a higher score must be preferred.

Not all the tweets which are being fetched are related to traffic. In this case, 48 tweets were actually seen to have been related to traffic, as shown in Figure 12.14. The total and traffic-related tweets count are shown in pie charts and bar plots, respectively, in Figures 12.15 and 12.16.

The relation or proportion between total tweets fetched and actual classified-traffic-tweets are depicted using various types of plots. The AUROC values are actually the area under the curves which have been plotted. This plot helps in comparing the accuracies of various algorithms, as shown in Figure 12.17. For the test case of "Seattle" shown in Figure 12.18, the user can interpret that they should avoid going through "Tacoma" on July 27, 2022, as there has been a roadblock.

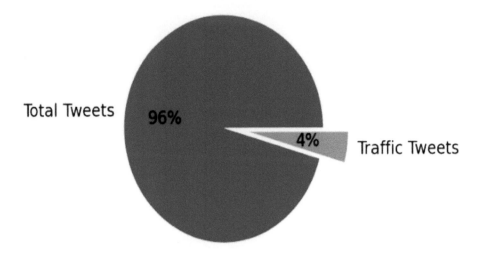

Figure 12.15   Pie chart (sample test).

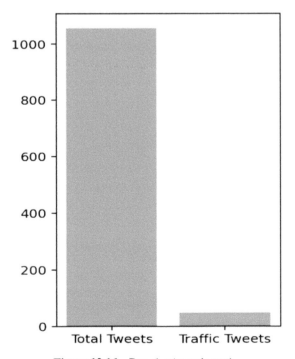

Figure 12.16   Bar plot (sample test).

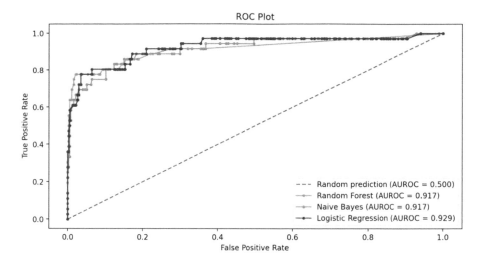

Figure 12.17 ROC plot and AUROC: sample test.

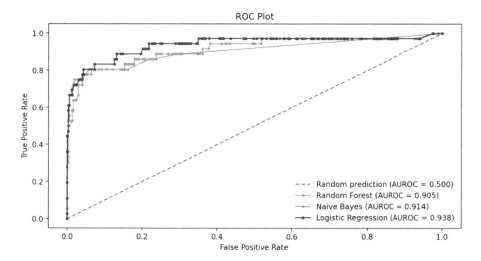

Figure 12.18 ROC plot for "Seattle."

The ROC curve for the city of "Seattle" using count vectorizer showed that logistic regression had the greatest AUROC, hence, the best accuracy. On clicking the "download the tweet as CSV" button, a dialogue box appears which enables us to save the dataset as a CSV file in the device's memory as shown in Figure 12.19.

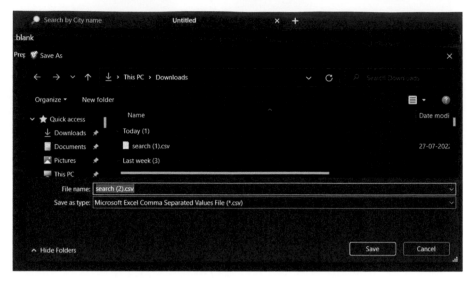

Figure 12.19   Downloading CSV dialogue box.

12.8.5 Deployment

The home page shown in Figure 12.20 serves as the landing page for the application. The general instructions for using the application, sample codes, and various links to documentation are available on this page. Various HTML and CSS properties have also been used in order to make the page attractive.

The contents of the page include:

- Overview of the app

- General instructions

- GitHub and drive links

- Sample codes for reference

- Documentation links

- Social media links

Figure 12.21 is the most important page of the app, where the user can have absolute versatility in obtaining tweets from Twitter API. One can set his own sets of parameters and choices for the prediction-classification model. The results of the search are displayed within the page itself, with the additional option of downloading the fetched tweets as a CSV file.

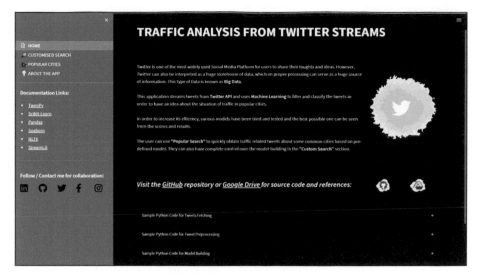

Figure 12.20   The home page.

The contents of the page include:

- General instructions.

- City name text input box.

- Slider to input the test–train split ratio.

- Dropdowns to select the vectorizer and machine learning algorithm to be used for the model.

- Search button to submit our choices.

The page shown in Figure 12.22 has been designed for time-saving and user accessibility purposes. Tweets can be fetched for the cities with just a single click of the button. All other functionalities, such as fetched tweets dataset, downloading the dataset, visualization, and ROC plots, are similar to that of the customized search page.

The following cities have been included:

- Los Angeles (USA)

- New York (USA)

- Seattle (USA)

- London (UK)

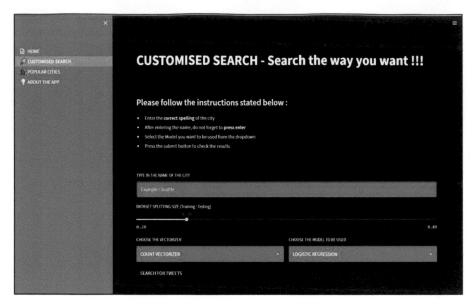

Figure 12.21   Customized search page.

Figure 12.22   Commonly searched cities page.

- Glasgow (UK)

- Manchester (UK)

- Ottawa (Canada)

- Toronto (Canada)

- Montreal (Canada)

The "About" page, as the name suggests, relays information about the app top the user about the developer and working of the app. For example, the libraries used for the deployment of the app are mentioned, and the application's workflow can be accessed for the user to understand how the application fetches and classifies tweets from the user.

The contents of the page include:

- User information

- Organization information

- Modules and libraries of Python utilized for this project

The data pipeline: The data pipeline primarily aims at preprocessing the traffic-tweets. The preprocessing tweets and machine learning workflow for classification.

The following tasks are intended to be included within the app:

- Increasing the number of machine learning algorithms being used in order to provide more versatility and wider scope for the users.

- Using various other types of vectorizers available.

- Development of a proper user registration and login system.

- Linking the application to a database.

- Development of an SMS-based or email-based alert system.

- Solve the problems regarding app deployment in the StreamLit cloud.

12.9 Conclusion

In this work, traffic-related tweets related to a particular city have been successfully mined from Twitter API. A proper pipeline was designed to preprocess tweets, followed by model building and machine learning to classify the fetched tweets. Machine learning algorithms and vectorization approaches were used, giving a wide variety of results. Versatility and efficiency were achieved by creating multiple approaches for model building. In the end, all the modules were integrated into a web application designed and deployed using StreamLit. As a result, the app was made versatile and interactive, enabling users to fetch traffic-related tweets from the city of their choice.

References

[1] Innamaa, S 2009, 'Self-adapting traffic flow status forecasts using clustering', IET Intelligent Transport Systems, vol. 3, no. 1, pp. 67–76.

[2] Williams, BM 2001, 'Multivariate Vehicular Traffic Flow Prediction Evaluation of ARIMAX Modeling', Transportation Research Board, vol. 1776, no. 1, pp. 194–200.

[3] Williams, B , Durvasula, P & Brown, D 1998, 'Urban freeway traffic flow prediction: application of seasonal autoregressive integrated moving average and exponential smoothing models', Transportation Research Record: Journal of Transportation Research Board, vol. 1644, no. 1, pp. 132–141.

[4] Williams, BM & Hoel, LA 2003, 'Modeling and forecasting vehicular traffic flow as a seasonal ARIMA process: theoretical basis and empirical results', Journal of Transportation Engineering, vol. 129, no. 6, pp. 664–672.

[5] Jayanthi. G, 'Multi Criteria Decision Making Analysis for sustainable Transport', Sustainability - Cases And Studies in using Operations Research and Management Science Methods, edited by Fausto Pedro Garcia Marquez and Benjamin Lev and to be published by Springer in 2022 (in press.)

[6] Jayanthi, G, Jothilakshmi, P: 'Traffic Time Series Forecasting on Highways – A Contemporary Survey of Models, Methods and Techniques', International Journal of Logistics Systems and Management, Inderscience, 2020.

[7] Jayanthi, G, Jothilakshmi, P: 'Prediction of traffic volume by mining traffic sequences using travel time based PrefixSpan', IET Intelligent Transport Systems, 2019, 13, (7),p. 1199–1210.

[8] Jayanthi G and Fausto Pedro García Márquez, 'Routing Vehicles on Highways by Augmenting Traffic Flow Network: A Review on Speed Up Techniques', Smart Innovation, Systems and Technologies, Vol. 273, 2021.

[9] Ermagun, A, Chatterjee, S & Levinson, D 2017, 'Using temporal detrending to observe the spatial correlation of traffic', PLoS ONE, vol. 12, no. 5, pp. 1–21.

[10] Ermagun, A & Levinson, DM 2018, 'Development and application of the network weight matrix to predict traffic flow for congested and uncongested conditions', Environment and Planning B: Urban Analytics and City, vol. 6, no. 9, pp. 1684–1705.

[11] Ermagun, A & Levinson, D 2018, 'Spatiotemporal traffic forecasting: review and proposed directions', Transport Reviews, vol. 38, no. 6, pp. 786–814.

[12] Zhuang, Y, Ke, R & Wang, Y 2019, 'Innovative method for traffic data imputation based on convolutional neural network', IET Intelligent Transport Systems, vol. 13, no. 4, pp. 605–613.

[13] Fournier-Viger, P, Lin, CW, Gomariz, A, Gueniche, T, Soltani, A, Deng, Z & Lam, HT, An Open Source Data Mining Library. Available from: <http://www.philippe-fournier-viger.com/spmf/>.[2016]

[14] Fournier-Viger, P, Usef, F , Roger, N & Nguifo, EM 2012, 'CMRules: Mining sequential rules common to several sequences', Knowledge Based System, vol. 25, no. 1, pp. 63–76.

[15] Ajantha Devi, V., & Nayyar, A. (2021). Evaluation of geotagging twitter data using sentiment analysis during COVID-19, Vol. 166.

[16] Subash Nadar, Ajantha Devi, Ruby Jain, Fadi Al-Turjman, 10 - Use of artificial intelligence in pharmacovigilance for social media network, Leveraging Artificial Intelligence in Global Epidemics, Academic Press, 2021, Pages 239–259, ISBN 9780323897778.

[17] Devi, V. A., & Naved, M. (2021). Dive in Deep Learning: Computer Vision, Natural Language Processing, and Signal Processing. In *Machine Learning in Signal Processing* (pp. 97–126). Chapman and Hall/CRC.

13

Harnessing the Power of Artificial Intelligence and Data Science

V. Sakthivel[1*], Devashree Pravakar[2], and P. Prakash[2]

[1]KADA, Konkuk University, Seoul, South Korea; Vellore Institute of Technology, India
[2]Vellore Institute of Technology, India

Email: mvsakthi@gmail.com; devashree.pravakar2020@vitstudent.ac.in; prakash.p@vit.ac.in

Abstract

Artificial intelligence (AI) and data science have a massive contribution by transforming the present world into a revolutionary step through their application in various fields to solve the most common problems. It gives the machines to exhibit human-like characteristics such as learning and problem-solving. Data science is the field of study of large amounts of data to discover previously unknown patterns and extract meaningful knowledge by combining domain expertise, using modern tools and techniques, and knowledge of mathematics and statistics. AI and data science are becoming much more important for the healthcare industry. It is helpful in getting better and faster diagnoses than humans. AI and data science applications in the domain of finance are huge. Time-series analysis and forecasting are useful tools for making quick and quality decisions that are for solving challenging real-time financial problems like stock market predictions. In today's world chatbots are used to make effective communication with its user. The use of chatbots is huge as they are universally used on various websites for providing information and guidance by interacting and explaining to human users how the company or product works with a quick and meaningful response. A virtual assistant is a digital assistant that recognizes simple voice commands and completes tasks for the user. This is also an application of AI and data science that enables virtual assistants to listen to the user's command, interpret it, and perform the task. It gives the users the power to set an alarm, play music from Spotify, make calls, send messages, make a shopping list, or provide information such as weather, facts from Wikipedia, search meaning of a word from the dictionary, etc., by browsing the web with just a simple natural language voice command. There are many more real-life applications, such as e-mail spam filtering, Recommendation systems, autocomplete, face recognition, etc.

13.1 Introduction

Understanding the relationship between artificial intelligence and data science:

The two technologies that are changing the world most rapidly right now are artificial intelligence and data science as shown in Figure 13.1. Data science is the use of massive volumes of data to anticipate the occurrence of future events by extracting, altering, and displaying the data. It examines the data's patterns and use visualization tools to create graphs that highlight the analytical steps. Then it creates prediction models to determine the possibility that future events will occur. GeeksforGeeks [1] says that in the area of data science, predictions are made using AI. Pre-processing, analysis, visualization, and prediction are all steps in the comprehensive process known as "data science." Artificial intelligence, on the other hand, uses a predictive model to foretell future events. AI uses computer algorithms, whereas data science uses a variety of statistical methodologies. This is due to the fact that data analysis and insight generation in data science need a number of phases. Data science is the study of hidden patterns in data and the creation of models based on statistical knowledge. Building models that mimic human understanding and cognition is the goal of artificial intelligence (AI). Compared to AI, data science does not require as much scientific processing.

With the aid of historical data [2], many tech companies are currently developing AI solutions that will enable them to enhance and automate difficult analytical tasks, study data in real time, modify their behavior with little need for supervision, and improve the accuracy and efficiency of their existing systems.

13.2 Different Domains in AI and Data Science

13.2.1 Healthcare

Healthcare sector is one of the most important domains where the use of data science and AI has brought revolutionizing results [4]. They have helped in

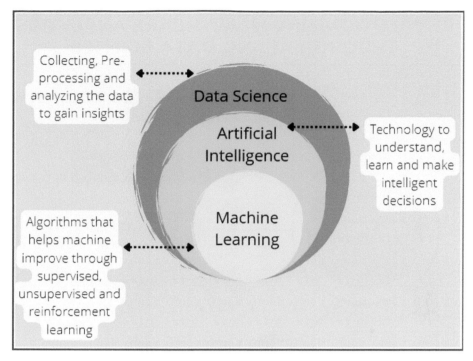

Figure 13.1 Comparison of various domains.

changing and enhancing the essential services in the healthcare industry. It gave the power to extract and study diverse healthcare data and help healthcare service providers to deliver the right treatment to the right patient at the earliest time and least cost.

Data in healthcare industry comes from various forms of resources as shown in Figure 13.2. Most of them comes from clinical notes, mobile devices, genomic sequences, medical images, wearable sensors, electronic health records (EHRs), social media, etc.

Following are the various applications of AI and data science.

13.2.1.1 Drug research

AI and data science help researchers to identify and develop a new drug, detecting its destructive potential in addition to its possible mechanisms of action [4]. As mentioned before, the data generated are huge in the healthcare industry. With the use of data science, evaluating the efficiency of the drug through data analysis became a much easier task.

Many diseases are emerging now and then, with the increase in climate change and pollution. This is also due to some other reasons like lack of proper food, physical illness, anxiety disorder, etc. Therefore, it has become

REVOLUTIONIZING HEALTHCARE ANALYTICS THROUGH AI AND DATA SCIENCE

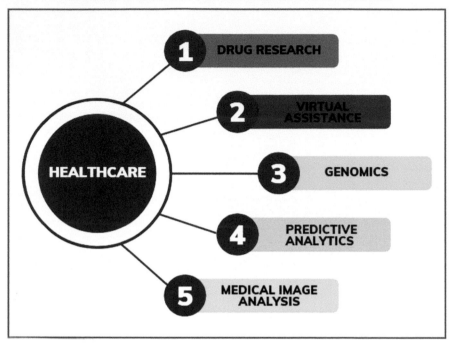

Figure 13.2 Healthcare applications.

really important for researchers to find the potential drug in a short period. Scientists have to do multiple tests to understand and identify the causative agent, and then find the formula and perform further tests to get the drug. In former times, it used to take 10–12 years, but now with the help of AI and data science, it is possible in lesser time.

AI techniques can also help a company reuse existing drugs and bio-active molecules. For drug companies, finding new applications for existing medications is a lucrative approach because it is feasible to remodel and reposition current medications than to develop them from scratch.

13.2.1.2 Virtual assistance

The application of virtual assistance is of great utilization of AI and data science. The virtual assistants can help in giving personalized experiences to the patients by identifying the disease by analyzing the symptoms [4]. Utilizing AI and data science, the medical platforms will assist in making recommendations for safety precautions, prescriptions, and necessary treatments based on the patient's health. The platform can also help in keeping track of the medicines and the treatment processes that are needed to follow and act as a reminder for regularly notifying the patients about proper medication, exercise, and food intake. This aids in preventing situations of carelessness that can aggravate the disease. Woebot, a virtual assistant created by Stanford University, is one example. It is a chatbot that assists people with psychological illnesses in getting the right care to improve their mental health.

Corti, an AI tool, aids emergency medical staff [5]. When a heart attack is suspected, Corti alerts emergency responders by listening to the background noise, the caller's speech, and relevant details from the patient's medical records. Like other ML systems, Corti does not search for specific signals but rather trains by listening to a lot of calls and picking out key components.

In light of this knowledge, Corti regularly enhances its model. The Corti system can differentiate between ambient sounds, such as sirens, caller-provided cues, or the background sound of the patient.

Artificial intelligence can help in analyzing verbal and non-verbal indications for the patient suffering from sudden cardiac arrest, and help them by establishing a diagnosis from a distance. This increases the chance of survival at a crucial time.

Around 73% of the time in Copenhagen, emergency dispatchers can determine a cardiac arrest from the caller's description. But AI is more capable. A 2019 small-scale study found that employing ML models could identify cardiac arrest calls more accurately than human dispatchers.

ML has the potential to be a crucial ally for emergency medical personnel. Upcoming forthcoming medicinal units may use technology to acknowledge to emergency calls using drone-mounted automatic defibrillators or volunteers who have been trained in CPR, increasing the chances of survival in cardiac arrest scenarios that occur in the community.

13.2.1.3 Genomics

Discovering the traits and anomalies in DNAs is the goal of genomics research [4]. Additionally, it aids in establishing a link between the illness, its symptoms, and the affected person's state of health. Additionally, the study of genomics examines how a particular type of DNA responds to drugs.

Prior to the development of effective data processing methods, studying genetics was a labor-intensive and time-consuming task. This is because the human body has millions of pairs of DNA cells. However, this process has recently become simpler because of the use of data science in the fields of healthcare and genetics. These techniques make it easier for researchers to identify particular genetic problems and the medication that works best for a particular type of gene.

Doctors can deliver the medication effectively if they are aware of how a patient's DNA cells react to a specific medicament. They can develop successful treatment plans to treat a condition for a specific patient, thanks to the beneficial insights into the genetic structure.

AI is also successfully utilized to hasten the discovery and development of new medical treatments. A change in molecular phenotypes, such as protein binding, favors genetic disorders. Predicting these changes entails estimating the risk that hereditary illnesses may manifest. Data on all discovered substances and biomarkers pertinent to particular clinical trials are gathered in order to achieve this. For instance, deep genomics' AI system processes this data. The company creates custom artificial intelligence (AI) and uses it to find new ways to correct genetic mutations' side effects while creating tailored treatments for people with rare Mendelian and complex diseases.

The company examines newly discovered compounds in order to generate genetic medicine more rapidly for conditions with significant unmet medical needs [7]. The team at the company is working on "Project Saturn," a drug system based on AI molecular biology that compares over 69 billion oligonucleotide molecules against 1 million target sites *in silico* (conducted or produced by means of computer modeling or computer simulation) in order to track cell biology and find more efficient treatments and therapies.

By lowering the expenses involved with treating rare diseases, the revelation and advancement of genetic medicine interests both patients and medical professionals.

13.2.1.4 Predictive analytics in healthcare

The predictive analytics model is based on AI and data science and produces predictions about the patient's status based on the information at hand [4]. The patient's body temperature, blood pressure, and sugar level are just a few examples of the data that can be included. The patient's data is gathered, and then it is examined to look for trends and connections. The goal of this method is to determine a disease's signs and symptoms, its stages, the degree of damage, and many other factors. Additionally, it aids in formulating plans for the proper course of treatment for the patient. Predictive analytics is therefore a very valuable method and it is very important to the healthcare sector.

Clinicians may improve their workflow, clinical judgment, and treatment plans by transforming electronic health records into an AI-driven prediction tool. A patient's whole medical history may be read in real time by

NLP and ML, which can then be linked to symptoms, ongoing conditions, or an ailment that affects other kins. The results can be used to develop a predictive analytics tool that can spot illnesses and cure them before they seriously affect human life.

In essence, the progression rate of chronic diseases may be predicted [8]. A business called CloudMedX specializes in interpreting notes-based, unstructured data (clinical notes, EHRs, discharge summaries, diagnosis and hospitalization notes, etc.).

Medical personnel can gain clinical insights from these notes and EHRs, enabling them to make data-driven decisions that will enhance patient outcomes. Applications of CloudMedX explanations have been previously made in a number of high-risk conditions, including renal failure, pneumonia, congestive heart failure, hypertension, diabetes, liver cancer, orthopedic surgery, and stroke, with the stated goal of reducing expenditures for outpatients and doctors by facilitating premature and precise patient diagnosis.

As a result, a sizable group of scientists and researchers with a variety of backgrounds – including artificial intelligence – have lately been mobilized to aid in the prediction of disease course and progression as well as the development of techniques to contain and manage COVID-19.

13.2.1.5 Medical image analysis

Analysis of medical images using image recognition technology is a futuristic approach [4]. Large amounts of data and photos from a clinical trial may need to be analyzed. Artificial intelligence (AI) algorithms can swiftly scan the datasets and compare them to the findings of other research to find patterns and unnoticed links. Medical imaging specialists can immediately track critical information, thanks to the technique. Data scientists are developing increasingly sophisticated methods to raise the standard of image analysis in order to effectively extract medical data from images.

Google AI has released a paper on utilizing deep learning to diagnose skin conditions. With a 97% accuracy rate, the deep learning model was developed to be able to identify 26 skin disorders. Data science, machine learning, and deep neural networks are used to make the diagnosis. Some of the common algorithms for medical picture analysis include the following:

- Anomaly detection algorithm: This algorithm aids in spotting abnormalities like bone displacements and fractures.

- Image processing algorithm: The denoising and enhancement of images is made possible with the aid of the image processing algorithm.

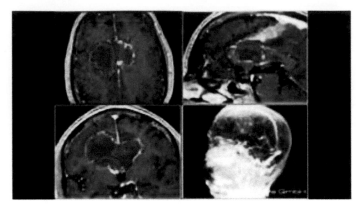

Figure 13.3   Tumor detection and calculation of the size of the abnormality. Source: Chimaera GmbH gallery.

Descriptive image recognition algorithm: It examines and makes use of the information it gathers from photographs and analyzes to create a more complete picture (for example, merging the images of the brain scan and designating them accordingly) as shown in Figure 13.3.

These strategies can be implemented by successfully using both supervised and unsupervised methods. AI is used to aid with case triage. A clinician's examination of images and scans is recommended. This enables radiologists or cardiologists to choose the data that is most important for prioritizing urgent situations, minimizing potential errors while reading electronic health records (EHRs), and generating more precise diagnoses.

As an illustration, Hardin Memorial Health (HMH) had to figure out how to extract pertinent data from EHRs in a concentrated form for imaging professionals [9]. With over 70,000 patients seen annually, the hospital's emergency room (ER) made the decision to work with IBM to put "The Patient Synopsis" into practice. With the help of this tool, users can locate patient data pertinent to the imaging process they underwent.

The Patient Synopsis gives radiologists and cardiologists a report that focuses on the context for these images by examining prior diagnostic and medical procedures, lab data, medical history, and known allergies. The product may be updated without affecting how the medical unit typically operates, and it can be accessed from any networked communication workstation or device.

The creation of a more specialized, focused, and accurate report used in the diagnostic decision-making process is made possible by identifying

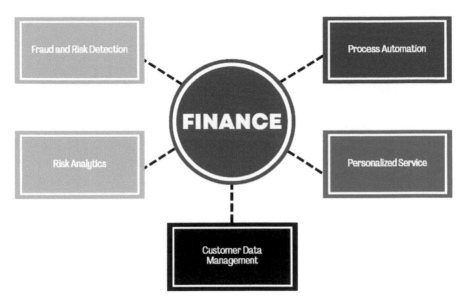

Figure 13.4 Applications of AI and data science in finance.

pertinent concerns and providing them to radiologists in a nice summarized view.

13.2.2 Finance

Data science and AI have many applications in the finance domain. The following are the most important applications as shown in Figure 13.4.

13.2.2.1 Fraud and risk detection

Online purchases are growing quickly in the digital age [10]. This is followed by several unethical actions elsewhere. For instance, there have been instances of fake insurance claims that have caused significant financial damage. Financial institutions have started implementing various solutions to address such serious issues. To keep track of consumer transactions and their prior history, the business software is coupled with data science tools.

The security systems of financial organizations have become incredibly secure and effective with the use of data science in finance. In addition, there has been less credit card fraud than in the previous five years of data. The software's algorithms are created in a way that makes it attempt to learn from prior data. Then, it makes predictions about potential dangers based on its learning experience. This prevents the financial companies from declining, thanks to AI and data science.

13.2.2.2 Risk analytics

Every firm faces risks related to security, rivals, financial loss, customer loss, and business failure for a variety of reasons [10]. Risk analytics becomes one of the crucial business processes, particularly for financial institutions like banks and insurance firms. Using risk analytics, unique business strategies can be developed to maintain a company's efficiency. Additionally, it contributes to the development of consumer and market trust.

Data science and machine learning are now used in the risk analytics operations. Data science in finance examines consumer and market patterns. Then, using machine learning methods, it tries to identify potential dangers. Risk analytics software aids in the development of company failure prevention measures. Organizations can also forecast business fluctuations brought on by numerous actions in the global market by studying the risks.

13.2.2.3 Customer data management

Data is one of the key resources in the corporate world, as was already established. Today, most businesses use customer data to increase efficiency. Financial firms use customer data to track consumer transactions. In the past, businesses saved data using conventional techniques. But the development of numerous technologies has given the finance sector a new focus [10].

Modern technologies used by the banking industry include big data tools, data science, artificial intelligence, and machine learning. Data management is now lot easier than it was before the creation of these technologies. Financial institutions are utilizing data science to efficiently handle and store the data of their customers. These institutions provide both structured and unstructured data storage solutions. However, data science techniques may be used to analyze, store, and segment any forms of data. The saved data may be used to produce the financial reports for these organizations.

13.2.2.4 Personalized services

In today's commercial world, personalized services are highly essential. The application of modern technology to customer service has allowed any company organization to accelerate its overall growth. These services provide applications that offer clients a customized experience using data science and artificial intelligence. Financial institutions have found this to be successful as well [10].

Data about customers' transactions, purchases, debits and credits, loan repayments, and bank balances can help businesses customize their customer offerings. For analyzing customer data, there may be other undisclosed factors. The application generates a credit score for each user based on this assessment. The bank determines which consumers are profitable based on credit scores. Then, it offers these clients deals, rebates, loans, and insurance that may be advantageous to them.

Customer service has also significantly improved, thanks to chatbots powered by AI and machine learning. Chatbots are used to deliver excellent customer support in the majority of financial industry sectors, including insurance businesses, loan providers, trading companies, banks, etc. These businesses are now utilizing speech recognition technology to increase client interaction. These and other AI and data science applications in finance are assisting companies in making large profits.

13.2.2.5 Process automation

Modern business executives arc seeking robotic process automation to streamline their operations and reduce costs. Intelligent character recognition makes it feasible to execute repetitive operations that formerly took a lot of time and effort. Software with AI capabilities can carry out operations including data extraction, document inspection, and data validation. Financial organizations significantly minimize the amount of labor-intensive manual work and fault rates by utilizing robotic process automation [12].

In the financial sector, data science and AI are quickly changing how we communicate and conduct business. The industry anticipates that the rise of blockchain technology will lead to better transaction prices and enhanced account security for customers. More advanced self-help and virtual reality technology that enhance the customer service experience are also something we may anticipate. Finally, financial organizations are expected to provide more individualized and rewarding services to their customer base in order to increase customer loyalty and gain a competitive edge. This is made possible by having more complete, accurate, and timely information about each client and their behavior and preferences.

13.2.3 E-commerce

AI and data science play a very important role in the E-commerce industry. There are many applications of these in this domain, such as recommendation systems, pricing algorithms, predictive segmentation, personalized product image search algorithm, and intelligent chatbots.

13.2.3.1 Recommendations system

With the help of AI and data science, recommendation systems can provide customers personalized list of products that they are most likely to purchase based on their prior buying behavior. When compared to search algorithms, recommendation systems are advantageous. They offer customized products according to the customers' tastes and assist consumers in finding items they might not have otherwise discovered. To make the user's purchasing experience more delightful by automating the search process, providing tailored items, and saving them time, any large platform needs a recommendation system algorithm [13]. AI recommendation systems combine data science and user data to filter and offer the best items to a particular user. The content recommendation system is compared to an experienced shop assistant who can offer more enticing products while also raising conversion rates since they are familiar with the user's wants, preferences, and requirements [17].

13.2.3.2 Pricing

The use of pricing algorithms to alter prices in response to data changes in the market or within the organization is another significant application of AI and data science in e-commerce [13]. Product pricing is not just a crucial duty for the producer or store; it also affects the customer, depending on rival prices. AI programs can create intelligent entry pricing, discounts, promotions, and other prices. Flexible pricing algorithms are created and trained by e-commerce businesses so that they may learn the necessary patterns and perform price optimization depending on them. These clever pricing algorithms can identify correlations in vast amounts of data that your company may want. Rapid response to market opportunities boosts return on investment.

13.2.3.3 Customer segmentation

Data science in e-commerce can be used to analyze and segment customer behavior. The use of artificial intelligence (AI) algorithms by e-commerce companies can estimate consumer interest in your products. Customer segmentation has been used to target marketing initiatives frequently [14]. Customers can be categorized into groups like loyal customers, high spenders, etc., based on their geographic, demographic, behavioral, and psychological commonalities.

Implementing customer segmentation creates a plethora of brand-new business opportunities [17]. There are several aspects that can be optimized, including budgeting, product design, promotion, marketing, and customer satisfaction. Simply said, the clustering technique is used to create customer segmentation models. Although there are various strategies, the *K*-means clustering technique is the most popular one. When the clusters are developed and focused, marketing initiatives are undertaken and high-producing clusters are discovered. Understanding the retail business for the commodity you are attempting to sell in detail is necessary to locate these clusters.

13.2.3.4 Virtual assistants

Intelligent virtual assistants, the next-generation chatbots, offer a better user experience [15]. Natural language processing is used by this AI-powered software to manage questions and provide responses that are more human-like.

Every e-commerce company needs to offer customer support. It goes without saying that a buyer would wish to allay any worries before making an online purchase. Nearly every e-commerce business offers a "chat now" option on their website for this purpose. Artificial intelligence has replaced the dedicated staff that was formerly assigned by corporations to handle live chat and customer support. The e-commerce company's AI chatbots will text or speak with customers online. By guiding customers through the vast array of products and their characteristics on the website, comparing products, and putting together combinations for customers, they will assist with personalized shopping.

Modern e-commerce users do not even realize that they are speaking to a machine and not a human because AI-based chatbots have advanced to such a degree. The advent of AI-powered chatbots has improved the overall quality of live chat by making it more personalized and intelligent, leading to greater customer experience.

13.2.3.5 Sentiment analysis

Even though practically every online company has a rating and review section on their website, it might be difficult to understand some reviews because they are posted by non-native speakers of the language. Such evaluations include misspellings, abbreviations, and slang. Such reviews can be analyzed by data scientists by utilizing NLP techniques, which are common in data science.

Data scientists further separate the reviews they have retrieved and do a customer sentiment analysis on them. This allows them to better understand why a particular review was negative. To improve consumer happiness, e-commerce businesses can use this data to reassess and plan their product and service offerings [16].

The easiest and most accessible instrument for an analyst to conduct customer sentiment analysis is social media. He/she uses language processing to find terms that reflect the attitude of the buyer toward the brand, either positively or negatively. This feedback enables the company to better serve customers' demands by improving products and services. Here also AI and data science play a very important role.

13.2.4 Education

AI and data science is rapidly transforming and improving the way the education industry operates. However, when data science and artificial intelligence are used correctly, personalized learning experiences can be created, which may aid in the resolution of some of these issues.

There is a tremendous amount of information gathered and retained on pupils' internet behavior and grades [18]. Since there is so much data available, AI algorithms can now offer experiences that were not possible before. A significant data trail is left behind when students use instructional content from internet sources. As a result, recommendations, learning routes, and assistance can be generated using this data while feeding AI models.

In this digital age, students are faced with an overwhelming amount of resources. It can be difficult to locate the specific documents that will answer their inquiry, and it is quite simple to lose interest in the process. Organizations are integrating chatbots into their learning platforms with the aid of natural language processing, an area of AI. The ability of these chatbots to find information that answers students' queries has been demonstrated. It has been demonstrated that the learning process is enhanced by using this approach of information retrieval rather than manually looking for the solution.

When a student watches a video, joins a class, is inactive, submits an assignment, etc., information is created every time. The information obtained from this can then be used to make predictions about the routes most likely to result in student success. By identifying issues early enough to allow the proper corrective actions to be performed, AI learning systems would be assisting students in realizing their full potential and perhaps even preventing them from dropping out.

13.2.5 Transportation

The importance of AI and data science in transportation can be summed up as faster, more effective, less expensive, and cleaner transportation. Technologies and intelligent solutions are revolutionizing how we move people and commodities.

13.2.5.1 Predictive analysis

Transportation service providers can optimize delivery timetables and route planning by using AI-based predictive analytics. Additionally, the technology-based strategy guarantees enhanced asset performance due to prompt maintenance, which lowers the incidence of breakdowns [19].

AI may be used to analyze the current routing, optimize track routes, and enhance upcoming ones. AI-based systems can improve routes and cut shipping costs by using digital and satellite maps, traffic data, and other data.

AI-driven solutions automatically collect and analyze asset data, and they will alert you to potential failures. These systems can compile data on failures, create statistics, and plan repairs based on this statistical information.

AI is able to learn from human judgments, form opinions, and communicate with people [20]. These tools can be used to distribute data to the appropriate database, fill out some invoices and web forms, and automate access to shipping data. Data derived from prior experiences, AI can also assist in predicting demand or enhance demand forecasting and warehouse management, with the aid of which a thorough examination of all the variables that could affect demand can yield insightful information.

13.2.5.2 Traffic management

In order to improve traffic management, AI and data science tools can take into account all the information that is currently available regarding vehicle speed, weather and road conditions, special events, and even accidents. This is made possible by a combination of real-time data, historical trends, and intelligent algorithms. These systems can be used by local governments and logistics corporations to improve traffic management, relieve congestion, and reduce CO_2 emissions [20].

An intelligent traffic management system (ITMS) uses artificial intelligence in conjunction with cameras placed at traffic junctions to find and identify cars breaking traffic regulations as well as to provide real-time notifications at the traffic control center. The law allows for the automated imposition of penalties and the electronic notification of offenders. Furthermore, based on the information acquired from these cameras, an adaptive traffic-control system may adjust traffic signal cycles in real time in response to shifting traffic conditions. When fully operational, the ATCS technology will cut down traffic light wait times by around half [21].

Combining modern communication technologies with artificial intelligence (AI) might help us manage our crowded streets so that they can

Figure 13.5   Pedestrian detector tutorial. Source: Omdena.com.

accommodate the growing number of automobiles by analyzing massive volumes of data in real time.

13.2.5.3 Pedestrian detection

Pedestrian detection will enable computer systems to detect and identify pedestrians in photos and videos automatically as shown in Figure 13.5. A system like this would undoubtedly aid driverless vehicles in avoiding hazardous situations, potentially greatly reducing the number of traffic accidents [22]. Detecting pedestrians is actually a significant challenge in pattern recognition and computer vision since they can be highly unexpected in the context of moving automobiles. Being so unexpected, they pose one of the largest risks to the development of self-driving cars. The most important thing is that a system can correctly distinguish a human from another object and comprehend what a pedestrian is going to do next rather than only being able to recognize individual human traits like beards and noses. Computer vision systems use bounding boxes as a starting point for the process of detecting and visualizing pedestrians. To recognize pedestrians, a variety of feature types have been used, including motion-based features, texture-based features, shape-based features, and gradient-based features.

Human pose estimation, a method that compiles data on a specific human's immediate behavior, has also been used in some ways. This is intended to tell the autonomous vehicle of the next move the pedestrian intends to make.

13.2.5.4 Supply chain management

In order to manage assets and facilities in a way that is as cost-effective as possible, AI is the greatest assistant for global supply chains.

Global logistics firms strive to maintain the stability of their supply chains, and using AI and data science to do so is not only feasible but also efficient [20].

AI and data science can help with route optimization, delivery sequences, and delivery status. Additionally, you are no longer limited to using only historical data. With considerably more accuracy now, product demand and consumer needs may be predicted. In other words, big data opens the door to improved resource allocation and supply chain optimization. This reduces the possibility of the alleged "dead stock" problem by allowing distributors and wholesalers to forecast the appropriate number of products they will be able to sell.

13.2.5.5 Airline route planning

The airline industry experiences significant losses on a global scale. Companies are having difficulty maintaining their occupancy ratios and operational earnings, with the exception of a few aviation service providers. The requirement to provide clients significant discounts and the sharp increase in the cost of aviation fuel has made the situation much worse. Airlines firms finally started utilizing data science to pinpoint important areas for improvement [23].

The airline firms can now estimate flight delays using AI and data science, choose the type of aircraft to purchase, and determine whether to arrive immediately at the destination or make a stop along the way. An airplane may travel directly from New Delhi to New York, for instance. As an alternative, it can decide to stop in any nation and successfully implement client loyalty programs.

13.2.6 Agriculture

AI and data science in agriculture have aided farmers in gaining a deeper understanding of farming data and procedures. Better judgments are made as a result, as are increased output and efficiency. AI is being used more and more in agriculture to track labor on farms, insect and weed activity, crop health, and soil health.

13.2.6.1 Controlling wild animal attacks

In a rural agricultural location, there is now less chance that domestic and wild animals may accidentally ruin crops or break in and steal things [26]. Thanks to the rapid advancements in video analytics enabled by AI algorithms, everyone interested in farming can secure the perimeters of their crops and structures. Systems for video surveillance using artificial intelligence and machine learning can scale for both small-scale farms and industrial-scale agricultural operations. Surveillance systems that use machine learning can be programmed or educated over time to distinguish between people and vehicles.

13.2.6.2 Forecast yield rates

It is feasible to estimate the prospective yield rates of a specific field before a vegetation period even starts. Using a mix of machine learning algorithms to analyze 3D mapping, social condition data from sensors, and soil color data from drones, agronomists can now forecast the prospective soil yields for a specific crop. A number of flights are conducted to get the most precise dataset possible [24].

By utilizing real-time sensor data and data from drone visual analytics, AI and machine learning enhance agricultural production forecast. Agricultural professionals now have access to entirely new datasets due to the volume of information being acquired by clever sensors and drones transmitting live video. It is now viable to integrate in-ground sensor data on moisture, fertilizer, and natural nutrient levels to analyze crop development trends across time [25]. Machine learning is the best method for integrating huge datasets and providing constraint-based suggestions for maximizing agricultural yields. Following is an illustration of how artificial intelligence, machine learning, underground sensors, infrared imaging, and real-time video analytics all combine to give farmers new information about how to improve crop health and yields as shown in Figure 13.6.

13.2.6.3 Price estimation

Accurate price estimates for crops based on yield rates are necessary to develop pricing strategies for a specific crop since they help calculate total quantities produced. By comprehending crop output rates and quality levels, agricultural firms, co-ops, and farmers may better bargain for the greatest possible price for their harvests. Whether a crop's price elasticity curve is inelastic, unitary, or extremely elastic – and takes into consideration the crop's overall demand – determines the pricing strategy. By alone, knowing this knowledge helps agricultural businesses avoid losing millions of dollars in revenue every year [24].

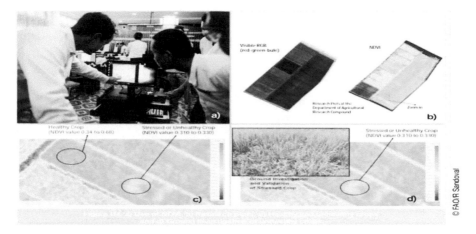

Figure 13.6 (a) Use of NDVI. (b)Research plots. (c) Healthy and unhealthy crops. (d) Ground investigation of unhealthy crops.

13.2.6.4 Livestock care

One of the booming applications of AI and data science in agriculture is the monitoring of cattle health, including vital signs, daily activity levels, and food consumption.

In these respects, artificial intelligence (AI) and data science are crucial to the agricultural sector [24]. It is crucial to understand what makes cows satisfied and happy daily using AI and machine learning in order to increase milk production. This industry offers completely new insights on how many farms that depend on cows and other livestock may be more successful.

13.2.6.5 Weather prediction

Weather greatly affects agricultural production and has an impact on the development, growth, and yield of crops. Crops can sustain physical harm from weather anomalies, and soil can erode.

Weather affects how well crops grow and mature before they are sold. The quality of the crop might be negatively impacted by bad weather while it is being transported or stored. Data science professionals are skilled in the use of techniques that reveal links and patterns that could otherwise go undetected. Through the investigation of particular elements causing weather change, they can reach conclusions that advance agricultural research [24].

13.2.6.6 Disease detection and pest management

Advanced algorithms are employed in modern agriculture to recognize natural patterns and behaviors that aid in forecasting the invasion of pests and

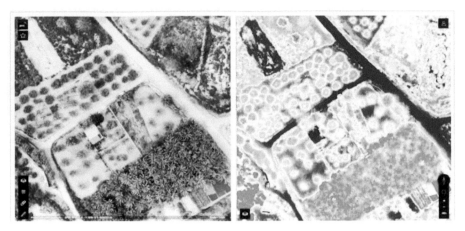

Figure 13.7 Monitoring the palms health.

the spread of microscopic illnesses as shown in Figure 13.7. Modern agricultural analytics are guiding how farmers should control pests. In agriculture, harmful insects are being dealt with scientifically using digital tools and data analysis [26].

Agricultural pests can swiftly deplete a farmer's revenues [24]. But improper pesticide use can harm humans, plants, and other living things. Some insects, however, can be extremely helpful to farmers and their crops, while others can be poisonous and transmit diseases. By deploying drones to photograph the field and then analyzing those photos to find diseased regions, disease detection can be accomplished.

13.3 Conclusion

This chapter discusses some of the most widespread real-world uses of artificial intelligence and data science in the contemporary advanced world across several sectors. It would take a long time to enumerate all the potential applications for these technologies in AI, and there are countless more. The relationship between data science and artificial intelligence was also covered in this chapter. Additionally, it gave a comparison of the two technologies and explored the functions of artificial intelligence in data science. By giving data science cutting-edge tools for appropriate predictive analysis and proper parameters for data engineering to be applied to software as well, artificial intelligence plays a significant role in the field.

Acknowlegdements

I am grateful to my parents, friends, and family for their encouragement during the course of the work for the support.

I really value the considerate feedback provided by the Books & Texts anonymous peer reviewers. This study has benefited greatly from everyone's goodwill and wisdom, and I have avoided making many errors because of it.

References

[1] GeeksforGeeks. (2022, July 12). Difference Between Data Science and Artificial Intelligence. https://www.geeksforgeeks.org/major-applications-of-data-science/

[2] GeeksforGeeks. (2021, March 12). Major Applications of Data Science. https://www.geeksforgeeks.org/major-applications-of-data-science/

[3] Mohammed, A. (2022, July 6). Artificial Intelligence in Data Science: 5 Definitive Facts. Learn | Hevo. https://hevodata.com/learn/artificial-intelligence-in-data-science/

[4] Intellipaat, S. (2022b, May 18). Application of Data Science in Healthcare. Intellipaat Blog. https://intellipaat.com/blog/data-science-applications-in-healthcare/

[5] Stig Nikolaj Blomberg, Fredrik Folke, Annette Kjær Ersbøll, Helle Collatz Christensen, Christian Torp-Pedersen, Michael R. Sayre, Catherine R. Counts, Freddy K. Lippert, Machine learning as a supportive tool to recognize cardiac arrest in emergency calls, Resuscitation, Volume 138, 2019, Pages 322–329, ISSN 0300-9572, https://doi.org/10.1016/j.resuscitation.2019.01.015. (https://www.sciencedirect.com/science/article/pii/S0300957218309754)

[6] "Artificial Intelligence, Machine Learning and Genomics." Genome.Gov, 12 Jan. 2022, www.genome.gov/about-genomics/educational-resources/fact-sheets/artificial-intelligence-machine-learning-and-genomics.

[7] "Project Saturn | Deep Genomics." Deep Genomics, www.deepgenomics.com/project-saturn. Accessed 27 Sept. 2022.

[8] "Healthcare Data Platform with Aggregation, Automation and AI." CloudMedx, 22 Sept. 2022, cloudmedxhealth.com.

[9] "Hardin Memorial Health | IBM." Hardin Memorial Health | IBM, www.ibm.com/case-studies/hardin-memorial-health-watson-health. Accessed 27 Sept. 2022.

[10] Intellipaat, S. (2022, January 22). Top Application of Data Science in Finance Industry in 2022. Intellipaat Blog. https://intellipaat.com/blog/data-science-applications-finance/

[11] Team, CRN. "Role of Data Science and AI in Customer Service Management - CRN - India." CRN - India, 23 Mar. 2022, www.crn.in/columns/role-of-data-science-and-ai-in-customer-service-management.

[12] A. (2022, May 4). The Role of Data Science and AI in Financial Decisions. Intellect Data. https://intellectdata.com/the-role-of-data-science-and-ai-in-financial-decisions/

[13] GeeksforGeeks. (2020, October 6). Top Applications of Data Science in E-commerce. https://www.geeksforgeeks.org/top-applications-of-data-science-in-e-commerce/

[14] Kumar, Dhiraj. "Implementing Customer Segmentation Using Machine Learning [Beginners Guide] - Neptune.Ai." Neptune.Ai, 18 June 2021, neptune.ai/blog/customer-segmentation-using-machine-learning.

[15] Mori, Margherita. (2021). AI-Powered Virtual Assistants in the Realms of Banking and Financial Services. 10.5772/intechopen.95813.

[16] Lin, Xiaoxin. (2020). Sentiment Analysis of E-commerce Customer Reviews Based on Natural Language Processing. 32–36. 10.1145/3436286.3436293.

[17] Vedala, R. (2022, July 21). Data Science and Machine Learning in the E-Commerce Industry: Insider Talks About Tools, Use-Cases, Problems, and More. Neptune.Ai. https://neptune.ai/blog/data-science-and-machine-learning-in-the-e-commerce-industry-tools-use-cases-problems

[18] Adke, Aarti. "Importance of Data Science and Artificial Intelligence in Education Sector." Importance of Data Science and Artificial Intelligence in Education Sector, 23 Apr. 2022, www.thehansindia.com/hans/young-hans/importance-of-data-science-and-artificial-intelligence-in-education-sector-739514.

[19] Abduljabbar, Rusul & Dia, Hussein & Liyanage, Sohani & Bagloee, Saeed. (2019). Applications of Artificial Intelligence in Transport: An Overview. Sustainability. 11. 189. 10.3390/su11010189.

[20] Haponik, A. (2022, May 25). AI, big data, and machine learning in transportation. Addepto. https://addepto.com/ai-big-data-and-machine-learning-in-transportation/

[21] Kumar, Vipul. "Artificial Intelligence: Application in Traffic Management System | by Vipul Kumar | Medium." Medium, 14 Oct. 2018, medium.com/@wrandomwriter/artificial-intelligence-application-in-traffic-management-system-9b5e3d7e620c.

[22] Guo, Lie & Ge, Ping-Shu & Zhang, Mingheng & Li, Linhui & Zhao, Yibing. (2012). Pedestrian detection for intelligent transportation systems combining AdaBoost algorithm and support vector machine. Expert Syst. Appl.. 39. 4274–4286. 10.1016/j.eswa.2011.09.106.

[23] "How Airlines Use Artificial Intelligence and Data Science in Operations | AltexSoft." AltexSoft, 25 Nov. 2021, www.altexsoft.com/blog/engineering/ai-airlines.

[24] Columbus, Louis. "10 Ways AI Has The Potential To Improve Agriculture In 2021." Forbes, 17 Feb. 2021, www.forbes.com/sites/louiscolumbus/2021/02/17/10-ways-ai-has-the-potential-to-improve-agriculture-in-2021/?sh=cff38227f3b1.

[25] Chandrasekar Vuppalapati "Machine Learning and Artificial Intelligence for Agricultural Economics" https://doi.org/10.1007/978-3-030-77485-1

[26] Srichandan, Pratik & Mishra, Ashis & Singh, Harkishen. (2018). Data Science and Analytic Technology in Agriculture. International Journal of Computer Applications. 179. 21–28. 10.5120/ijca2018916850.

14

Determining the Severity of Diabetic Retinopathy through Neural Network Models

Kaushik Ghosh[1], Sugandha Sharma[1], Tanupriya Choudhury[1,2], and Husain Falih Mahdi[3]

[1]UPES, Bodholi Campus, India.
[2]CSE Department, Symbiosis Institute of Technology, Symbiosis International (Deemed University), Pune, India.
[3]Department of Computer and Software Engineering, University of Diyala Baquba, Iraq

Email: k79aushik@gmail.com; sugandhasahrma2016@gmail.com; tanupriya@dd.upes.ac.in; tanupriya.choudhury@sitpune.edu.in; hussain.mahdi@ieee.org

Abstract

Diabetic retinopathy is a disease caused in human beings as a result of diabetic complications. The light-sensitive tissues present in human eye are responsible for vision. Diabetic retinopathy in its severe form can damage the blood vessels of these tissues, causing permanent loss of vision to a patient. A considerable number of people across the globe are susceptible to this problem. The worst part of the problem is that the patients may remain asymptomatic at the initial stages of the disease. However, detection of the disease at an early stage can save a patient from losing his/her vision. One of the techniques of detecting the stage of this disease is through the study of retinal images. This chapter presents a comprehensive comparative analysis of different neural network models, viz. VGG16, InceptionV3, Resnet50, and MobileNetV; in detecting diabetic retinopathy by classifying the severity of the disease on a scale of 0–4. Retinal images with multiple properties like varying contrast, intensity, brightness, etc., were used to train the models and predict the stage of disease the patient is in at present. The results of the analysis reveal that the model MobileNetV2 records maximum training and testing accuracy amongst the rest. Moreover, accuracy graph, loss graph, confusion matrix, and classification report for every model have also been presented, for increasing number of epochs. It can be seen that the overall performance of all the models improves with an increase in the number of epochs.

14.1 Introduction

Diabetic retinopathy (DR) occurs when the retina of the eye gets affected as a result of diabetes [1]. As per the reports, 0.8 million people have completely lost their eyesight and another 3.7 million were affected visually around the globe in 2010 alone [2, 3]. During the period of 2015–2019, a global prevalence of 27% was found for the disease DR [4, 5]. Although the patients remain asymptomatic at the early stages of this disease, yet a prolonged ailing might even lead to complete loss of eyesight [6]. But it is to be noted that even before the vision gets affected, DR in itself can reach a fairly advanced stage in the patient [4]. In 2015, around 4.24 million U.S. population over the age of 40 were impacted due to irreversible impairment of vision, which included blindness. This figure is expected to increase to approximately 8.96 million by 2050 [7]. Looking at the trends, it was estimated that by 2030, approximately 191.0 million of DR patients are expected worldwide [4, 5]. As a result, regular checkups are an essential part of the treatment of diabetes, particularly for patients who have crossed the middle age. Having said that, it has been seen that a considerable number of patients with diabetes were not recommended for any annual eye examination [8–10]. DR happens due to high glucose levels in the blood, harming the little veins of retina [11]. The outcome of which is additional liquid, cholesterol, blood, alongside different fats in the retina, bringing about thickening and growing of the macula. In the cycle of providing test blood, the retina begins developing some new irregular fragile veins called IrMAs (intraretinal microvascular abnormalities) [12]. The enlarged weight in the eye may bring about harming the optic nerve in the later stages of the disease. DR causes harm to retinal veins leading to loss of vision. Retinal screening is a potential answer for detecting the harm caused to the retina during the initial stages [11]. DR exhibits no symptom at first and thus the greater part of the patients stays unaware of the condition unless the disease affects their vision. Therefore, early screening and identification of the disease is the key to prevent and control any further advancement in the stages of it. A primary indication of DR is the presence of exudates, which might be visible in the fundus image of the eyes [13]. Optical coherence tomography (OCT) and images of fundus are two major sources for detecting and assessing the retinal condition, by ophthalmologists worldwide [13]. These images are fed into different deep learning models, including convolutional neural networks (CNNs), for analyzing and extracting unique features for classifying and analyzing different stages of DR, which, if unattended, may eventually cause vision loss.

Therefore, in this chapter, we have collected over 3500 images of fundus and have used them to identify the stage of the disease present in a patient, using deep learning techniques.

Diabetic retinopathy usually advances through the following four stages:

1. Mild non-proliferative retinopathy: This is the earliest stage and is identified with the presence of miniature aneurysms (MA) [14].

2. Moderate non-proliferative retinopathy: In this stage, the retinal veins may twist and expand as the disease advances from stage 1. This might cause the veins to lose their capacity for blood transportation [14].

3. Severe non-proliferative retinopathy: In this stage, numerous veins are blocked, which, in turn, denies blood flow into the retina and brings about denied blood flexibly to retina because of the blockage of more number of veins [14].

4. Proliferative diabetic retinopathy: In this stage, the blood vessels of retina are severely damaged. As a result, the retina generates new and non-robust blood vessels, which are not normal. Moreover, they grow on the retinal surface and thereby disrupt blood supply in the retina .The new blood vessels cause vitreous hemorrhage due to leakage. The exudates of the hemorrhage block the light rays from reaching the retina, leading to dark perception and loss of vision.

The assessment of the seriousness and level of retinopathy related to an individual having diabetes is currently performed by clinical specialists. This process is dependent on the retinal pictures of the eyes of the patient. With the increase in the number of diabetic patients, the volume of the retinal image dataset keeps on increasing every passing day. This increases the workload of the clinical specialists manifold. Therefore, to provide a clinically effective and cost-effective detection of DR, we require an automated optimized deep learning classification model, which will identify the severity of the diabetic disease in the patient.

The rest of the chapter has been organized as follows. Section 14.2 contains the background of the topic, Section 14.3 contains discussion about the different models used for classification, Section 14.4 presents the results obtained through experiments, and, finally, Section 14.5 contains the drawn inferences and conclusions.

14.2 Background

Although a great deal of literature is available at present on the application of machine learning and deep learning in the medical field, yet not many relevant works were found for DR in particular.

The authors in [15] proposed a computer-aided detection mechanism to identify the anomalies in the retinal fundus images. This work is focused on image data augmentation and extracted features and classified different stages of DR, using machine learning techniques.

Image analysis techniques were used in [16] for automated and timely detection of DR. A similar approach was proposed in [17] for identification of various stages of the disease. The authors performed static wavelet transforms for the retinal fundus images along with CLAHE (contrast limited adaptive histogram equalization) for the enhancement of vessels.

A novel technique based on saliency to detect leakages in the fluorescence angiography was proposed in [18].

A hybrid classifier-based approach was taken in [19] to detect the retinal lesions. Classification was done by pre-processing the extracted lesions and formulation of features.

Karen Simonyan and Andrew Zisserman performed a thorough evaluation of CNNs by increasing the network depth and studying its effects on model accuracy [20].

The authors in [21] classified 1.2 million images of high resolution into 1000 separate classes using deep CNN while training the model. Top-1 and top-5 error rates of 37.5% and 17.0% were obtained, respectively, on the test data.

In [22], better performance was achieved by attention-based CNN over attention-based LSTM for answer selection.

The work in [23] advocated that a fine-tuned gated CNN is capable of modeling long context dependency.

The authors in [24] professed that deeper neural networks require more complex training techniques. A framework based on residual learning was presented to reduce the complexity of training deeper neural networks.

Machine learning techniques play a major role in detection as well as classification of DR [25]. An automated diagnosis method using SVM classifier was proposed in [25] for identifying different stages of DR, viz. normal, mild DR, moderate DR, severe DR, and prolific DR. The proposed method gave an accuracy of 82%. The same authors developed another system for identifying exudates and blood vessels in the raw images and SVM was used for classification [26]. This second system achieved a considerably high accuracy rate of 85.9%.

A CNN model was developed for identifying non-DR, NPDR, and PDR in [27]. In this work, the authors applied morphological processing techniques along with texture analysis upon the fundus images for detecting hard exudates and blood vessels. The model proposed by them achieved a classification accuracy of 93%. Another model for identifying features such as micro-aneurysms, exudates, and hemorrhages was proposed in [28]. This CNN model used data augmentation and was capable of identifying and differentiating between five different stages of DR. The model under consideration achieved an accuracy of 75%. In [29], the authors proposed a CNN for identifying accurately four different stages of DR, viz. non-DR, NPDR, severe NPDR, and PDR. This model attained a fairly high accuracy of 80.2%. A fine-tuned VGG16 model was proposed to classify images and achieved an accuracy of 77% [30].

A robust dataset of DR fundus images was developed in [31]. The authors used the images to train an improvisation of the InceptionV3 network. Their proposed model attained 88.72% accuracy for classifying the stages of DR. An architecture of a deep neural network for detecting and identifying the stage of retinopathy was proposed in [32]. This proposed model composed of a pair of sub-networks. The first one extracted high-level features from the images, which in turn were aggregated and was used as input for the second sub-network, in order to predict the severity of the disease. It gave an improved testing accuracy as compared to VGG16 and ResNet50.

CAD-based technique was proposed to screen the fundus images for micro-aneurysms in [33]. Soft and hard exudates were detected in an automated manner and the preliminary signs of DR were identified through an automatic approach [34]. The work in [35] detected macular anomalies on fundus images using filters, morphological operations and thresholds for the diagnosis of DR. In [36], the authors proposed an algorithm for the extraction of retinal vasculature in order for blood vessel detection.

14.3 Algorithms under Consideration

The four algorithms discussed in this chapter for classification of the images are: (i) VGG16 (without Gaussian blur), (ii) MobileNetV2, (iii) Resnet50, and (iv) InceptionV3. In this section, the architecture and working of each of these models have been discussed in a nutshell.

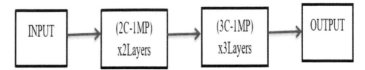

Figure 14.1 Architecture of VGG16 (without Gaussian blur).

14.3.1 VGG16 (without Gaussian blur)

VGG16 is a CNN architecture deliberated to be one of the exceptional vision model architecture till date. VGG16 has a large number of hyper-parameters and is focused on having convolution layers, each having 3×3 filter with a stride 1. It uses padding and maxpool layer of the 2×2 filter of stride 2. It follows this organization of convolution and maxpool layers regularly throughout the complete architecture. At the end, it has two FC (fully connected) layers succeeded by a softmax function for output that predicts a multinomial probability distribution in multiclass classification problems. VGG16 got its name due to its 16 layers with weights. This big network has parameters, which is approximately 138 million.

The architecture of VGG16 is given in Figure 14.1. The input to first layer is of fixed size 224×224 RGB image. The image is passed through a stack of subsequent layers, where the filters of size 3×3 are used. In one of the configurations, it also utilizes 1×1 convolution filters, which can be seen as a linear transformation of the input channels. The size of stride is usually 1, and the spatial padding usually used is 1 pixel for 3×3 convolutional layers. Spatial pooling is performed by five maxpooling layers, after passing through some of the convolutional layers. Maxpooling is performed over a 2×2 pixel window, with stride size of 2. A stack of convolutional layers (which typically have varied depths in various architectures) is followed by three fully connected (FC) layers. The first two layers have 4096 channels each, and the third conducts a 1000-way ILSVRC classification and hence comprises 1K channels (one channel per class). The last layer is the softmax layer. Configuration of FC layers is kept identical in all networks. All hidden layers are equipped with the ReLU. It has to be considered that many a times, none of the networks contain local response normalization (LRN). This kind of normalization fails to bring in any qualitative improvement on the performance on the ILSVRC dataset, but rather results in problems like enhanced consumption of memory as well as total computation time.

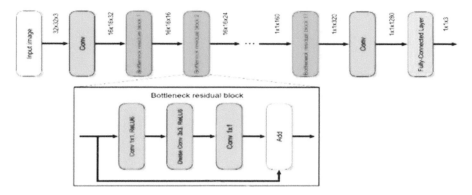

Figure 14.2   Architecture of MobilenetV2.

14.3.2 MobileNetV2

The ImageNet dataset was used to train this model. It is built on a residual structure that is inverted and the residual connections are sandwiched between the bottleneck layers. The intermediary expansion layer leverages depth-wise convolutions that are lightweight so as to sieve features as a source of non-linearity. This architecture (depicted in Figure 14.2) constitutes the initial fully convolution layer using 32 filters, succeeded by 19 residual bottleneck layers. Impeccable results were obtained for object detection as well as semantic segmentation while MobileNetV2 was used as the foundation for the purpose of extracting features.

There are two different types of blocks: first, a residual block having stride of 1 and, second, a block having stride of 2 used for downscaling. As can be seen from Figure 14.2 [40], there are three layers for each of the two types; the primary layer is a 1 × 1 convolution along with ReLU6. The subsequent layer is the depth-wise convolution. The final layer is additional 1 × 1 convolution but scans any non-linearity. It is asserted that if ReLU is used again, the deep neural networks merely have the advantage of a linear classifier that too on the non-zero volume part of the output domain. There is also an expansion factor known as the value of t, which has been set to 6 for all major experiments.

This implies that with the input of 64 channels, the internal output would be 384 channels. Usually, the principal network (width multiplier 1, 224 × 224) incurs a computational cost of approximately 300 million multiply-adds and utilizes 3.4 million parameters.

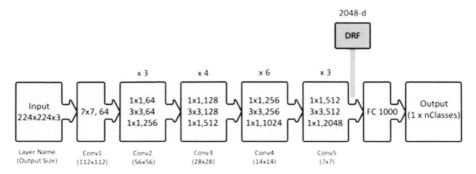

Figure 14.3 Architecture of ResNet50.

14.3.3 ResNet50

ResNet stands for residual networks. The numeral 50 in the algorithm name indicates the presence of 50 layers in the neural network.

ResNet50 architecture [39] is shown in Figure 14.3. A convolution having 64 different kernels each of size 7×7 and each having a stride that is of size 2 generates 1 layer. After that, we go for maxpooling with the same size of stride. In the succeeding convolution, we have a 1×1 convolution for 64 kernels. Following this, there is a 3×3 convolution, with 64 kernels, and, finally, a 1×1256 kernel. Subsequently, these three layers are then repeated thrice so as to give us a total of nine layers in this particular step. This is again succeeded by a kernel of 1×1128, then a kernel of 3×3128, and lastly a kernel of 1×1512. These steps were repeated in four number of times so as to generate 12 layers in this particular step. This whole setup is trailed by a kernel of 1×1256 and two kernels with 3×3256 and one kernel of 1×11024. This step is reiterated six times to generate 18 layers in total. Further, a 1×1512 kernel with two more of 3×3512 and $1 \times 12,048$ kernels are present. Repeating this three times provides us nine more layers. Hereafter, we go for an average pooling and conclude it with a totally connected layer comprising 1000 nodes. A softmax function is applied at the end to give a single layer.

This entire process generates 50 layers deep convolutional network excluding activation functions and the max/or average pooling layers.

Discrete residual flow (DRF) that has been removed from the last convolutional layer of this network has also been shown. K in the figure represents the filter that is of size K. n represents the number of channels. Fully connected layer of 1000 neurons is indicated by FC 1000. The numerals present on the top of each block represent the recurrence of each block. nClasses denotes the total instances of output classes.

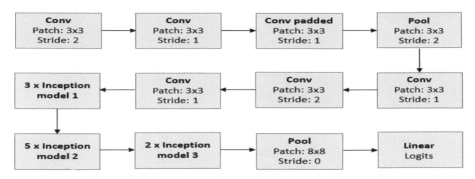

Figure 14.4 Architecture of InceptionV3.

14.3.4 InceptionV3

Google's InceptionV3 is the third version in a series of deep learning CNN architectures. It is 48 layers deep. These algorithms are computationally more efficient in terms of two major aspects: (i) parameters generated and (ii) the cost overhead with respect to resources like memory [37].

The architecture of InceptionV3 (Figure 14.4) [38] network is progressively built in the following steps: (i) factorized convolutions to decrease the computational efficacy by reducing the number of parameters used in a network. This is followed by (ii) smaller convolutions. This is achieved by replacing convolutions of larger size with convolutions of smaller size. This replacement leads to rapid training. For example, a 5 × 5 filter takes 25 parameters, whereas two 3 × 3 filters that replace a 5 × 5 filter will have 18 parameters only. The smaller convolutions are further followed by (iii) asymmetric convolutions. Asymmetric convolution can be explained as a 3 × 3 convolution substituted by a 1 × 3 convolution. Then we have a 3 × 1 convolution. The number of parameters in an asymmetric convolution is however smaller as compared to symmetric convolutions.

After asymmetric convolution come the (iv) auxiliary classifiers. It is a small CNN implanted between the layers during the phase of training and it acts as a regularizer. The final step is (v) reduction of grid size, which is performed by pooling operations.

14.4 Data Collection

A total of 3648 high-resolution fundus images were taken for experimentation purpose. The images of fundus utilized in this chapter are available publicly in Kaggle [41]. The images were provided by Asia Pacific Tele-Ophthalmology Society. The link for the dataset is given below:

https://www.kaggle.com/c/aptos2019-blindness-detection/data

```
In [1]: import tensorflow as tf
        import tensorflow.keras as tfk
        import tensorflow_hub as th
        import pandas as pd
        import datetime
        import numpy as np
        import matplotlib.pyplot as plt
        import os
        import math
        import cv2
        from sklearn.metrics import confusion_matrix, classification_report
        import seaborn as sns

In [2]: Images = np.load("/content/drive/MyDrive/Major Project/9-GB-Dataset/Complete-9-GB-Dataset.npz")["arr_0"]

In [3]: Labels = pd.read_csv("/content/drive/MyDrive/Major Project/9-GB-Dataset/train.csv")
```

Figure 14.5 (a)   Code snippet for loading dataset.

```
In [5]: Labels = Labels.diagnosis.values

In [7]: # Plotting the Images
        fig, ax = plt.subplots(5, 5, figsize = (20, 15))
        ax = ax.ravel()
        for i in np.arange(0, 25):
            ax[i].imshow(Images[i])
            ax[i].set_title("Severity: {}".format(Labels[i]))
            ax[i].axis("off")

        plt.subplots_adjust(hspace = 0.3, wspace = 0.3)
```

Figure 14.5(b)   Code snippet for plotting images.

Figure14.5(a) and Figure 14.5(b) represent loading of dataset and plotting of the images.

14.5 Implementation Steps of the Algorithms

This section provides the implementation steps for the four algorithms under consideration along with code snippets for the same.

14.5.1 Implementation steps of VGG16

1.   Sequential method is used. Sequential model means that all the layers of the model will be arranged in sequence.

2. VGG16 model has been applied on a set of images with Gaussian blur and another time on a set of images without Gaussian blur.

3. An object of ImageDataGenerator for both training and testing data is used.

4. Added the layers and also pretrained weights of the model.

5. In the dense layer, the "softmax" activation function is used.

6. Adam optimizer is used to obtain the global minima while training our model.

7. Set the metrics on accuracy.

8. Model is built successfully.

    Figure 14.5(c), 14.5(d), 14.5(e), 14.5(f), 14.5(g) and 14.5(h) represent the code snippets for implementing VGG16.

14.5.2 Implementation of the MobileNetV2 model

1. Sequential model is used.

2. Layers have been added along with pretrained weights of the model by providing the path of the URL to download it from the internet.

3. Trainable argument for each layer has been disabled explicitly.

4. 1024 units of neurons with activation function as "ReLU" in the dense layer has been used.

5. In the output layer, the "softmax" activation function is used.

6. The loss function selected is SparseCategorialCrossentropy and the optimizer is Adam.

7. While compiling the model, the matrix has been set to accuracy.

8. Model is built.

    Figure 14.6(a), 14.6(b) and14.6(c) represent the code snippets for implementing MobileNetV2.

14.5.3 Implementation steps of ResNet50

1. Sequential model is used.

```
In [8]: Labels

Out[8]: array([2, 4, 1, ..., 2, 0, 2])

In [9]: GB_Images = Images.copy()

In [10]: GB_Images.shape

Out[10]: (3662, 224, 224, 3)

In [11]: # Implementing the Gaussian Blur
         for i in range(GB_Images.shape[0]):
             GB_Images[i] = cv2.GaussianBlur(GB_Images[i], (0, 0), 15)

In [12]: # Plotting the Gaussian Blurred Images
         fig, ax = plt.subplots(5, 5, figsize = (20, 15))
         ax = ax.ravel()
         for i in np.arange(0, 25):
             ax[i].imshow(GB_Images[i])
             ax[i].set_title("Severity: {}".format(Labels[i]))
             ax[i].axis("off")

         plt.subplots_adjust(hspace = 0.3, wspace = 0.3)
```

Figure 14.5(c)　Code snippet for implementing Gaussian blur for loading dataset.

Implementing VGG Normal (Without Gaussain Blur)

```
In [13]: inputShape = (224, 224, 3)

In [77]: VGG_Model = tfk.applications.vgg16.VGG16(input_shape = inputShape, weights = "imagenet", include_top = False)
         VGG_Model.trainable = False

In [78]: VGGModel = tfk.layers.Flatten()(VGG_Model.output)
         VGGModel = tfk.layers.Dense(units = 5, activation = "softmax")(VGGModel)
         VGGModel = tfk.Model(inputs = VGG_Model.input, outputs = VGGModel)
         VGGModel.summary()
```

Figure 14.5(d)　Code snippet for implementing VGG16 without Gaussian blur.

2.　Layers have been added along with pretrained weights of the model by providing the path of the URL to download it from the internet.

3.　Trainable argument for each layer has been disabled explicitly.

4.　1024 units of neurons with activation function as "ReLU" in the dense layer has been used.

```
In [17]:  # Creating Funcitons for the Model Metrics Plotting & Model Testing
          def plotConfusionMatrix(y_true, y_pred, cmap = None):
              plt.style.use("seaborn-poster")
              plt.figure(figsize = (15, 9))
              sns.heatmap(confusion_matrix(y_true, y_pred), annot = True, cbar = False, cmap = cmap, fmt = "g")
              plt.xticks(np.arange(0.5, 5.5, 1), ["No DR", "Mild", "Moderate", "Severe", "Proliferate DR"])
              plt.yticks(np.arange(0.5, 5.5, 1), ["No DR", "Mild", "Moderate", "Severe", "Proliferate DR"], rotation = 45)
              plt.show()

          def plotClassificationReport(y_true, y_pred):
              print(classification_report(y_true, y_pred))

          def plot_Graphs(trainedModel):
              plt.style.use("seaborn-poster")
              plt.figure(figsize = (15, 7))
              plt.plot(trainedModel.history["accuracy"], label = "Accuracy", linewidth = 5)
              plt.plot(trainedModel.history["val_accuracy"], label = "val_Accuracy", linewidth = 5)
              plt.title("Accuracy Graph")
              plt.xlabel("Epochs")
              plt.ylabel("Accuracy")
              plt.legend()
              plt.show()

              plt.figure(figsize = (15, 7))
              plt.plot(trainedModel.history["loss"], label = "Loss", linewidth = 5)
              plt.plot(trainedModel.history["val_loss"], label = "val_Loss", linewidth = 5)
              plt.title("Loss/Cost Graph")
              plt.xlabel("Epochs")
              plt.ylabel("Loss")
              plt.legend()
              plt.show()
```

Figure 14.5(e)   Code snippet for model metrics plotting and model training (VGG16).

```
In [82]:  trainLimit = 2300
          validationLimit = 3000

          trainedModel1 = VGGModel.fit_generator(
                              Augmented_Data.flow(Images[:trainLimit],
                                                  Labels[:trainLimit],
                                                  batch_size=batchSize),
                              verbose = 1,
                              validation_data = ValidationData.flow(Images[trainLimit: validationLimit],
                                                                    Labels[trainLimit:validationLimit],
                                                                    batch_size = batchSize),
                              validation_freq = 1,
                              steps_per_epoch = math.ceil(2300 // batchSize),
                              validation_steps = math.ceil(700 // batchSize),
                              callbacks=[earlyStopping, modelCheckpoint, Tensorboard_Checkpoint_Callback],
                              epochs = 100)
```

Figure 14.5(f)   Code snippet for fit generator (VGG16).

```
In [10]: batchSize = 32
         Augmented_Data = tfk.preprocessing.image.ImageDataGenerator(rotation_range = 57,
                                                 width_shift_range=0.1,
                                                 height_shift_range=0.1,
                                                 horizontal_flip=True,
                                                 shear_range=0.1,
                                                 zoom_range=0.1)

         ValidationData = tfk.preprocessing.image.ImageDataGenerator()

In [80]: earlyStopping = tfk.callbacks.EarlyStopping(monitor = "val_accuracy",
                                               patience = 25,
                                               verbose = 1,
                                               mode = "max"
                                               )
         modelCheckpoint = tfk.callbacks.ModelCheckpoint(
                             "VGG_Normal.h5",
                             verbose = 1,
                             mode = "max",
                             save_best_only = True,
                             monitor = "val_accuracy"
                             )

         Tensorboard_Checkpoint_Callback = tfk.callbacks.TensorBoard(
                                 os.path.join("/content/drive/MyDrive/Major Project/9-GB-Dataset/New_logs/",
                                     datetime.datetime.now().strftime("%Y%m%d-%H%M%S"))
         )
```

Figure 14.5(g)   Code snippet for data augmentation and callback (VGG16).

VGG on Gaussian Blurred Images

```
In [07]: VGG_Model_GB = tfk.applications.vgg16.VGG16(input_shape = inputShape, weights = "imagenet", include_top = False)
         VGG_Model_GB.trainable = False

In [ ]: VGGModelGB = tfk.layers.Flatten()(VGG_Model_GB.output)
         VGGModelGB = tfk.layers.Dense(units = 5, activation = "softmax")(VGGModelGB)
         VGGModelGB = tfk.Model(inputs = VGG_Model.input, outputs = VGGModel)
         VGGModel.summary()

In [10]: VGGModel.compile(optimizer = "adam", metrics = ["accuracy"], loss = "sparse_categorical_crossentropy")

In [11]: trainedModel2 = VGGModelGB.fit_generator(
                             Augmented_Data.flow(GB_Images[:trainLimit],
                                 Labels[:trainLimit],
                                 batch_size=batchSize),
                             verbose = 1,
                             validation_data = ValidationData.flow(GB_Images[trainLimit:validationLimit],
                                 Labels[trainLimit:validationLimit],
                                 batch_size = batchSize),
                             validation_freq = 1,
                             steps_per_epoch = math.ceil(2100 // batchSize),
                             validation_steps = math.ceil(700 // batchSize),
                             callbacks=[earlyStopping, modelCheckpoint, Tensorboard_Checkpoint_Callback],
                             epochs = 100)
```

Figure 14.5(h)   Code snippet compilation (VGG16).

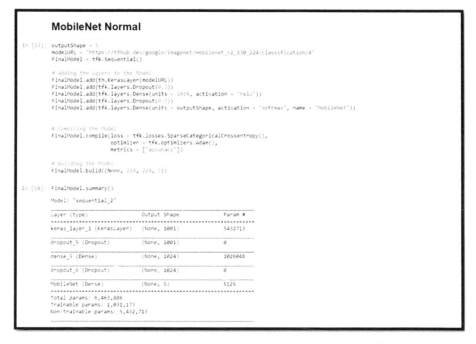

Figure 14.6(a)   Code snippet for adding layers (MobileNetV2).

5. In the output layer, the "softmax" activation function is used.

6. The loss function selected is SparseCategorialCrossentropy and the optimizer is Adam.

7. While compiling the model, the matrix has been set to accuracy.

8. Model is built.

Figure 14.7(a) and 14.7(b) represent the code snippets for implementing ResNet50.

14.5.4 Implementation of InceptionV3

1. Sequential model is used.

2. Layers have been added along with pretrained weights of the model by providing the path of the URL to download it from the internet.

3. 2048 units of neurons with activation function as "ReLU" in the dense layer has been used.

```
In [55]:  earlyStopping = tfk.callbacks.EarlyStopping(monitor = "val_accuracy",
                                                      patience = 25,
                                                      verbose = 1,
                                                      mode = "max"
                                                      )
          modelCheckpoint = tfk.callbacks.ModelCheckpoint(
                               "MobileNetv2_Normal.h5",
                               verbose = 1,
                               mode = "max",
                               save_best_only = True,
                               monitor = "val_accuracy"
                               )
          Tensorboard_Checkpoint_Callback = tfk.callbacks.TensorBoard(
                               os.path.join("/content/drive/MyDrive/Major Project/9-GB-Dataset/New_logs/",
                               datetime.datetime.now().strftime("%Y%m%d-%H%M%S"))
                               )

In [16]:  trainedModel3 = FinalModel.fit(
                               Augmented_Data.flow(Images[:trainLimit],
                                                   Labels[:trainLimit],
                                                   batch_size=batchSize),
                               verbose = 1,
                               validation_data = ValidationData.flow(Images[trainLimit:validationLimit],
                                                   Labels[trainLimit:validationLimit],
                                                   batch_size = batchSize),
                               validation_freq = 1,
                               steps_per_epoch = math.ceil(2300 // batchSize),
                               validation_steps = math.ceil(700 // batchSize),
                               callbacks=[earlyStopping, modelCheckpoint, Tensorboard_Checkpoint_Callback],
                               epochs = 100)
```

Figure 14.6(b)   Code snippet for callback (MobileNetV2).

```
In [61]:  from sklearn.metrics import accuracy_score

In [62]:  # Testing Accuracy
          accuracy_score(Labels[validationLimit:], y_pred)

Out[62]:  0.783987915407855

In [63]:  # Training Accuracy
          accuracy_score(Labels[:trainLimit], FinalModel.predict_classes(Images[:trainLimit]))

Out[63]:  0.8391304347826087
```

Figure 14.6(c)   Code snippet for testing and training accuracy (MobileNetV2).

4.   In the output layer, the "softmax" activation function is used.

5.   The loss function selected is SparseCategorialCrossentropy and the optimizer is Adam.

6.   While compiling the model, the matrix has been set to accuracy.

7.   Model is built.

Figure 14.8(a), 14.8(b) and 14.8(c) represent the code snippets for implementing InceptionV3.

14.6 Results

The accuracy graph, the loss graph, confusion matrix, and the classification report for each of the algorithms are given as follows.

These findings are further followed by a tabular representation of the training and testing accuracies of each of the said algorithms.

```
In [93]: %tensorboard --logdir "/content/drive/MyDrive/Major Project/9-GB-Dataset/New_logs/"

          <IPython.core.display.Javascript object>
```

ResNet

```
In [88]: resnetURL = "https://tfhub.dev/google/imagenet/resnet_v2_50/classification/4"

In [90]: # Creating a Sequential Model
          FinalModelRESNET = tfk.Sequential()

          # Adding the Layers to the Model
          FinalModelRESNET.add(th.KerasLayer(resnetURL))
          FinalModelRESNET.add(tfk.layers.Dropout(0.3))
          FinalModelRESNET.add(tfk.layers.Dense(units = 1024, activation = "relu"))
          FinalModelRESNET.add(tfk.layers.Dropout(0.3))
          FinalModelRESNET.add(tfk.layers.Dense(units = outputShape, activation = "softmax", name = "ResNet"))

          # Compiling the Model
          FinalModelRESNET.compile(loss = tfk.losses.SparseCategoricalCrossentropy(),
                                   optimizer = tfk.optimizers.Adam(),
                                   metrics = ["accuracy"])

          # Building the Model
          FinalModelRESNET.build((None, 224, 224, 3))
```

Figure 14.7(a)   Code snippet for adding layers (ResNet50).

```
In [91]: FinalModelRESNET.summary()

          Model: "sequential_4"

          Layer (type)                 Output Shape              Param #
          ======================================================================
          keras_layer_2 (KerasLayer)   (None, 1001)              25615849
          _____
          dropout_7 (Dropout)          (None, 1001)              0
          _____
          dense_7 (Dense)              (None, 1024)              1026048
          _____
          dropout_8 (Dropout)          (None, 1024)              0
          _____
          ResNet (Dense)               (None, 5)                 5125
          ======================================================================
          Total params: 26,647,022
          Trainable params: 1,031,173
          Non-trainable params: 25,615,849
          _____

In [92]: earlyStopping = tfk.callbacks.EarlyStopping(monitor = "val_accuracy",
                                                      patience = 25,
                                                      verbose = 1,
                                                      mode = "max"
                                                      )
          modelCheckpoint = tfk.callbacks.ModelCheckpoint(
                                "ResNet_Normal.h5",
                                verbose = 1,
                                mode = "max",
                                save_best_only = True,
                                monitor = "val_accuracy"
                                )

          Tensorboard_Checkpoint_Callback = tfk.callbacks.TensorBoard(
                                os.path.join("/content/drive/MyDrive/Major Project/9-GB-Dataset/New_logs/",
                                datetime.datetime.now().strftime("%Y%m%d-%H%M%S"))
                                )
```

Figure 14.7(b)   Code snippet for callback (ResNet50).

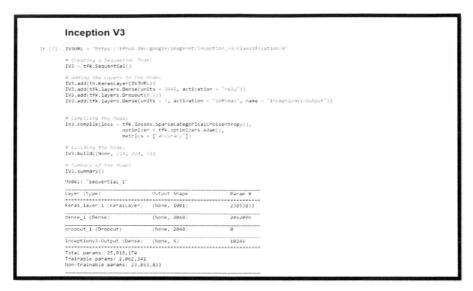

Figure 14.8(a)   Code snippet for the sequential model (InceptionV3).

The accuracy graphs have epochs on the *x*-axis and accuracy on the *y*-axis. Similarly, the loss graphs are with epochs on the *x*-axis and loss on the *y*-axis.

A confusion matrix is used for measuring performance in machine learning algorithms and neural networks. It assists in visualizing a classifier's accuracy by performing comparison between the actual and predicted classes. It is composed of squares that represent true positive, true negative, false positive, and false negative.

Each column of confusion matrix represents the instances of the predicted class. However, each row will represent the instances of the actual class. The aggregation of correct predictions for a class falls in the predicted column and expected row for that class value. The aggregation of incorrect predictions for a class, on the other hand, falls into the expected row for that class value and the predicted column for that specific class value.

The images have been classified into five classes, viz. no diabetic retinopathy (No DR), mild diabetic retinopathy, moderate diabetic retinopathy, severe diabetic retinopathy, and proliferate diabetic retinopathy (proliferate DR) on the basis of severity.

Some sample fundus images from the dataset are given in Figure 14.9 [42]. The leftmost column of images belongs to the class No DR. The

```
In [8]: earlyStopping = tfk.callbacks.EarlyStopping(monitor = "val_accuracy",
                                                     patience = 25,
                                                     verbose = 1,
                                                     mode = "max"
                                                     )
        modelCheckpoint = tfk.callbacks.ModelCheckpoint(
                            "InceptionV3-Normal.h5",
                            verbose = 1,
                            mode = "max",
                            save_best_only = True,
                            monitor = "val_accuracy"
                            )

        Tensorboard_Checkpoint_Callback = tfk.callbacks.TensorBoard(
                            os.path.join("/content/drive/MyDrive/Major Project/9-GB-Dataset/New_logs/",
                                         datetime.datetime.now().strftime("%Y%m%d-%H%M%S"))
                            )

In [12]: trainLimit = 2300
         validationLimit = 3000
         trainedModel5 = IV3.fit(
                             Augmented_Data.flow(Images[:trainLimit],
                                         Labels[:trainLimit],
                                         batch_size=batchSize),
                             verbose = 1,
                             validation_data = ValidationData.flow(Images[trainLimit: validationLimit],
                                                     Labels[trainLimit:validationLimit],
                                                     batch_size = batchSize),
                             validation_freq = 1,
                             steps_per_epoch = math.ceil(2300 // batchSize),
                             validation_steps = math.ceil(700 // batchSize),
                             callbacks=[earlyStopping, modelCheckpoint, Tensorboard_Checkpoint_Callback],
                             epochs = 100)
```

Figure 14.8(b)   Code snippet for callback (InceptionV3).

```
In [96]: trainedModel4 = FinalModelRESNET.fit(
                             Augmented_Data.flow(Images[:trainLimit],
                                         Labels[:trainLimit],
                                         batch_size=batchSize),
                             verbose = 1,
                             validation_data = ValidationData.flow(Images[trainLimit: validationLimit],
                                                     Labels[trainLimit:validationLimit],
                                                     batch_size = batchSize),
                             validation_freq = 1,
                             steps_per_epoch = math.ceil(2300 // batchSize),
                             validation_steps = math.ceil(700 // batchSize),
                             callbacks=[earlyStopping, modelCheckpoint, Tensorboard_Checkpoint_Callback],
                             epochs = 100)
```

Figure 14.8(c)   Code snippet for training model (InceptionV3).

middle one corresponds to mild and moderate NPDR classes, respectively. The rightmost column belongs to severe NPDR and PDR classes, respectively.

The classification report visualizer displays the precision, recall, *F*1, and support scores for the model. The accuracy graphs, loss graphs, confusion matrix, and classification reports for all the four models under consideration are given below in section 14.6.1. Figure 14.10(a), 14.10(b), 14.10(c), 14.10(d), 14.11(a), 14.11(b), 14.11(c), 14.11(d), 14.12(a), 14.12(b), 14.12(c), 14.12(d), 14.13(a), 14.13(b), 14.13(c) and 14.13(d) represent the results of the different algorithms implemented.

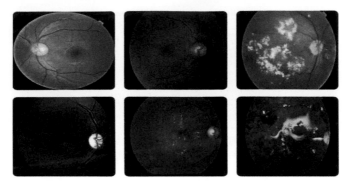

Figure 14.9   Fundus images for the five different stages of DR.

14.6.1 VGG16

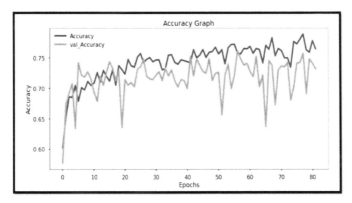

Figure 14.10(a)   Accuracy graph for VGG16.

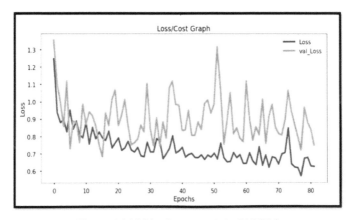

Figure 14.10(b)   Loss graph for VGG16.

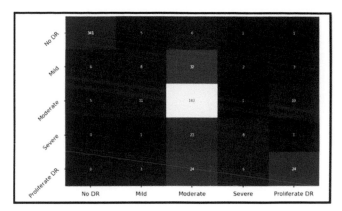

Figure 14.10(c) Confusion matrix for VGG16.

	precision	recall	f1-score	support
0	0.97	0.96	0.97	354
1	0.29	0.16	0.20	51
2	0.63	0.84	0.72	169
3	0.44	0.26	0.33	31
4	0.62	0.42	0.50	57
accuracy			0.79	662
macro avg	0.59	0.53	0.54	662
weighted avg	0.77	0.79	0.77	662

Figure 14.10(d) Classification report for VGG16.

14.6.2 MobileNetV2

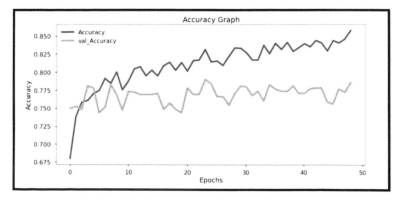

Figure 14.11(a) Accuracy graph for MobileNetV2.

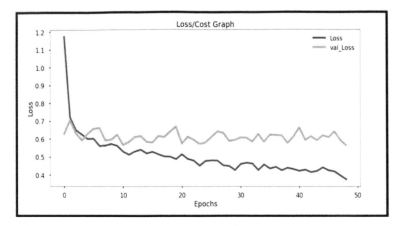

Figure 14.11(b) Loss graph for MobileNetV2.

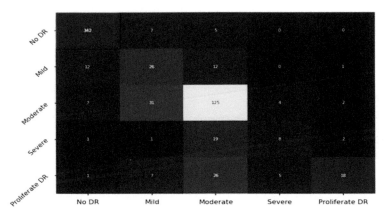

Figure 14.11(c) Confusion matrix for MobileNetV2.

```
              precision   recall  f1-score   support

           0       0.94     0.97      0.95       354
           1       0.36     0.51      0.42        51
           2       0.67     0.74      0.70       169
           3       0.47     0.26      0.33        31
           4       0.78     0.32      0.45        57

    accuracy                          0.78       662
   macro avg       0.64     0.56      0.57       662
weighted avg       0.79     0.78      0.78       662
```

Figure 14.11(d) Classification report for MobileNetV2.

14.6.3 ResNet50

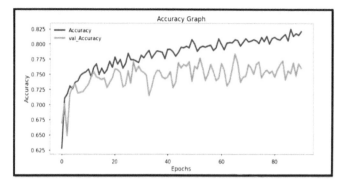

Figure 14.12(a) Accuracy graph for ResNet50.

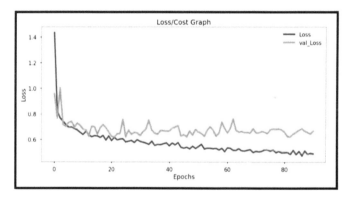

Figure 14.12(b) Loss graph for ResNet50.

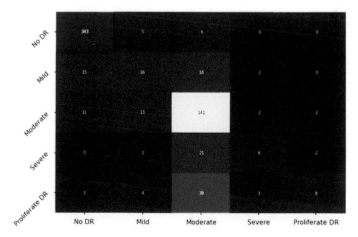

Figure 14.12(c) Confusion matrix for ResNet50.

	precision	recall	f1-score	support
0	0.92	0.97	0.94	354
1	0.40	0.31	0.35	51
2	0.63	0.83	0.72	169
3	0.46	0.19	0.27	31
4	0.67	0.14	0.23	57
accuracy			0.78	662
macro avg	0.62	0.49	0.50	662
weighted avg	0.76	0.78	0.75	662

Figure 14.12(d)   Classification report for ResNet50.

14.6.4 InceptionV3

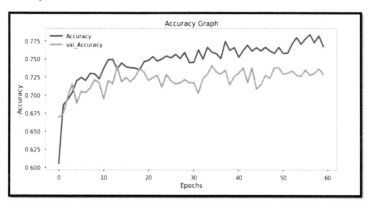

Figure 14.13(a)   Accuracy graph for InceptionV3.

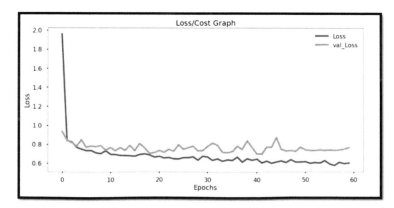

Figure 14.13(b)   Loss graph for InceptionV3.

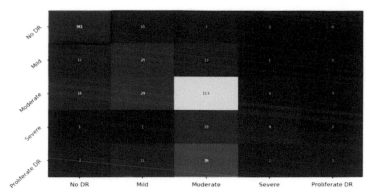

Figure 14.13(c)   Confusion matrix for InceptionV3.

```
              precision    recall  f1-score   support

           0       0.91      0.96      0.94       354
           1       0.33      0.49      0.39        51
           2       0.60      0.67      0.63       169
           3       0.47      0.26      0.33        31
           4       0.38      0.05      0.09        57

    accuracy                           0.74       662
   macro avg       0.54      0.49      0.48       662
weighted avg       0.72      0.74      0.72       662
```

Figure 14.13(d)   Classification report for InceptionV3.

Table 14.1   Comparison of testing and training accuracies.

Model name	Training accuracy	Testing accuracy
InceptionV3	~76%	~73%
ResNet50	~81%	~77%
MobileNetV2	**~85%**	**~79%**
VGG16 (without Gaussian blur)	~77%	~76%

It can be seen from Table 14.1 that MobileNetV2 gives maximum training accuracy and testing accuracy of 85% and 79%, respectively, as compared to the other models under consideration.

14.7 Conclusion

As it can be seen from the accuracy and loss graphs of different models, it is evident that the accuracy increases and loss decreases with increase in the number of epochs, for all the four models. Therefore, it may be concluded that the overall performance of the models gets better with the increase in the number of epochs. The training accuracy of InceptionV3, ResNet50, MobileNetV2, and VGG16 (without Gaussian blur), and MobileNetV2 were recorded as 76%, 81%, 85%, and 77%, respectively. Similarly, the testing accuracy of InceptionV3, ResNet50, MobileNetV2, and VGG16 (without Gaussian blur) and MobileNetV2 were recorded as 73%, 77%, 79%, and 76%, respectively.

So, as a final inference, MobileNetV2 may be considered to be the best model for classifying the images on the basis of severity of the disease diabetic retinopathy.

References

[1] Jen Hong Tan, Hamido Fujita, Sobha Sivaprasad, Sulatha V. Bhandary, A. Krishna Rao, Kuang Chua Chua and U. Rajendra Acharya "Automated segmentation of exudates, haemorrhages, microaneurysms using single convolutional neural network," *Information Sciences*, vol. 420, pp. 66–76, 2017.

[2] M. Mateen, J. Wen, Nasrullah, S. Song, and Z. Huang, "Fundus image classification using VGG-19 architecture with PCA and SVD," *Symmetry*, vol. 11, no. 1, 2019.

[3] Kangrok Oh,Hae Min Kang,Dawoon Leem,Hyungyu Lee,Kyoung Yul Seo and Sangchui Yoon, Early detection of diabetic retinopathy based on deep learning and ultra-wide-field fundusimages,https://www.nature.com/articles/s41598-021-81539-3,Article number: 1897 (2021).

[4] International Diabetes Federation. Diabetes atlas. *IDF Diabetes Atlas*, 9th edn. (International Diabetes Federation, Brussels, 2015).

[5] Ting, D. S. W., Cheung, G. C. M. & Wong, T. Y. Diabetic retinopathy: Global prevalence, major risk factors, screening practices and public health challenges: A review. *Clin. Exp. Ophthalmol.* 44, 260–277 (2016).

[6] Thomas, R., Halim, S., Gurudas, S., Sivaprasad, S. & Owens, D. Idf diabetes atlas: A review of studies utilising retinal photography on the global prevalence of diabetes related retinopathy between 2015 and 2018. *Diabetes Research and Clinical Practice*, p. 107840 (2019).

[7] Welp A, Woodbury RB and McCoy MA,Meeting the Challenge of Vision Loss in the United States: Improving Diagnosis, Rehabilitation, and Accessibility, https://www.ncbi.nlm.nih.gov/books/NBK402380/, 2016 Sep 15.

[8] Cynthia Owsley, Gerald McGwin, Kay Scilley, Christopher A Girkin, Janice M Phillips and Karen Searcey.Perceived barriers to care and attitudes about vision and eye care: Focus groups with older African Americans and eye care providers. Investig. Ophthalmol. Vis. Sci. 47, 2797–2802 (2006)

[9] MacLennan, P. A., McGwin, G., Searcey, K. & Owsley, C. A survey of Alabama eye care providers in 2010–2011. BMC Ophthalmol. 14, 44 (2014).

[10] Chiu-Fang Chou, Cheryl E Sherrod, Xinzhi Zhang, Lawrence E Barker, Kai McKeever Bullard, John E Crews and Jinan B Saaddine, Barriers to eye care among people aged 40 years and older with diagnosed diabetes, 2006–2010. Diabetes Care 37, 180–188 (2014).

[11] Christian Rask-Madsen and George L. King ,Vascular complications of diabetes: mechanisms of injury and protective factors, Cell Metab. 2013 Jan 8; 17(1): 20–33.

[12] Nihat Sayin, Necip Kara, and Gökhan Pekel,Ocular complications of diabetes mellitus, World J Diabetes, 2015 Feb 15; 6(1): 92–108.

[13] Praveen Vashist, Sameeksha Singh,Noopur Gupta and Rohit Saxena, Role of Early Screening for Diabetic Retinopathy in Patients with Diabetes Mellitus: An Overview, Indian J Community Med.2011 Oct-Dec; 36(4): 247–252.

[14] Ramandeep Singh, Kim Ramasamy,Chandran Abraham,Vishali Gupta and Amod Gupta, Diabetic retinopathy: An update, Indian J Ophthalmol 2008 May-Jun; 56(3): 179–188.

[15] V. Raman, P. Then and P. Sumari. "Proposed retinal abnormality detection and classification approach: Computer-aided detection for diabetic retinopathy by machine learning approaches", Communication Software and Networks (ICCSN), 2016 8th IEEE.

[16] N. Singh and R. Chandra Tripathi. "Automated early detection of diabetic retinopathy using image analysis techniques", International Journal of Computer Applications, vol. 8, no. 2, (2010), pp. 18–23.

[17] T. A. Soomro, "Role of Image Contras Enhancement Technique for Ophthalmologist as Diagnostic Tool for Diabetic Retinopathy", Digital Image Computing: Techniques and Applications (DICTA), 2016 International Conference on. IEEE, (2016).

[18] Y. Zhao, "Intensity and Compactness Enabled Saliency Estimation for Leakage Detection in Diabetic and Malarial Retinopathy", IEEE Transactions on Medical Imaging, vol. 36, no. 1, (2017), pp. 51–63.

[19] M. U. Akram, "Detection and classification of retinal lesions for grading of diabetic retinopathy", Computers in biology and medicine, vol. 45, (2014), pp. 161–171.

[20] Karen Simonyan and Andrew Zisserman, "Very deep convolutional networks for large- scale image recognition", 2014.

[21] Krizhevsky, A., Sutskever, I., Hinton, G.E.: ImageNet classification with deep convolutional neural networks. In: Advances in Neural Information Processing Systems, pp. 1097–1105 (2012).

[22] Ying Wen, Weinan Zhang, Rui Luo, and Jun Wang. 2016. Learning text representation using recurrent convolutional neural network with highway layers. SIGIR Workshop on Neural Information Retrieval.

[23] Yann N Dauphin, Angela Fan, Michael Auli, and David Grangier. 2016. Language modeling with gated convolutional networks. arXiv preprint arXiv:1612.08083 .

[24] He K. X., Ren S., Sun J.: Deep residual learning for image recognition.In:Proceedingd of the IEEE Conference on Computer Vision and Pattern Recognition ,pp.770–778(2016).

[25] Acharya U., Chua C., Ng E., Yu W., Chee C., Application of Higher Order Spectra for the Identification of Diabetes Retinopathy Stages, Journal of Medical Systems, vol. 32, no. 6, pp. 481–488, 2008.

[26] Acharya U., Lim C., Ng E., Chee C. and Tamura T., Computer-Based Detection of Diabetes Retinopathy Stages using Digital Fundus Images, Proceedings of the Institution of Mechanical Engineers, vol. 223, no. 5, pp. 545–553, 2009.

[27] Nayak J., Bhat P., Acharya R., Lim C. and Kagathi M., Automated Identification of Diabetic Retinopathy Stages using Digital Fundus Images, Journal of Medical Systems, vol. 32, no. 2, pp. 107–115, 2008.

[28] H. Pratt, F. Coenen, D. Broadbent, S. Harding, Y. Zheng, "Convolutional Neural Networks for Diabetic Retinopathy", International Conference on Medical Imaging Understanding and Analysis, Loughborough, UK, July 2016

[29] M. Shaban, Z. Ogur, A. Shalaby, A. Mahmoud, M. Ghazal, H. Sandhu, et al., "Automated Staging of Diabetic Retinopathy Using a 2D Convolutional Neural Network", IEEE International Symposium on Signal Processing and Information Technology, Louisville, Kentucky, USA, December 2018.

[30] Omar Dekhil, Ahmed Naglah, Mohamed Shaban, Ahmed Shalaby, Ayman El-Baz, "Deep-Learning Based Method for Computer Aided Diagnosis of Diabetic Retinopathy", IEEE International Conference on Imaging Systems & Techniques, Abu Dhabi, United Arab Emirates, December 2019.

[31] Gao Z., Li J., Guo J., Chen Y., Yi Z., and Zhong J., Diagnosis of Diabetic Retinopathy using Deep Neural Networks, IEEE Access Journal, vol. 7, pp. 3360–3370, 2018.

[32] Hu J., Chen Y., Zhong J., Ju R., and Yi Z., Automated Analysis for Retinopathy of Prematurity by Deep Neural Networks, IEEE Transactions on Medical Imaging, vol. 38, no. 1, pp. 269–279, 2019. pmid:30080144

[33] A. Mizutani, C. Muramatsu, Y. Hatanaka, S. Suemori, T. Hara and H. Fujita, "Automated Microaneurysm Detection Method Based on Double Ring Filter in Retinal Fundus Images", Proceedings of SPIE, 2009.

[34] H. Jaafar, A. Nandi and W. Al-Nuaimy, "Automated Detection of Exudates in Retinal Images using a Split-and-Merge Algorithm", 18th European Signal Processing Conference. Aalborg, Denmark, 2010.

[35] Pachiyappan A., Das U., Murthy T. and Tatavarti R., Automated Diagnosis of Diabetic Retinopathy and Glaucoma using Fundus and OCT Images, Lipids in Health and Disease, vol. 11, no. 73, 2012.

[36] Tan J., Acharya U., Chua K., Cheng C C. and Laude A., Automated Extraction of Retinal Vasculature, Medical Physics. Vol. 43, no. 5, pp. 2311–2322, 2016.

[37] https://blog.paperspace.com/popular-deep-learning-architectures-resnet-inceptionv3-squeezenet/

[38] Diagram Reference https://ieeexplore.ieee.org/document/8351550

[39] Diagram Reference https://www.mdpi.com/1424-8220/20/2/447

[40] DiagramRefcrence https://www.mdpi.com/1424-8220/20/14/3856

[41] Asia Pacific Tele-Ophthalmology Society, "APTOS 2019 blindness detection," Kaggle, https://www.kaggle.com/c/aptos2019-blindness-detection/data, 2019, [Dataset].

[42] Diagram Reference https://doi.org/10.1371/journal.pone.0233514.g001

15

Method for Muscle Activity Characterization using Wearable Devices

Wilver Auccahuasi[1], Oscar Linares[2], Karin Rojas[3], Edward Flores[4],
Nicanor Benítes[5], Aly Auccahuasi[6], Milner Liendo[7],
Julio Garcia-Rodriguez[8], Grisi Bernardo[9],
Morayma Campos-Sobrino[10], Alonso Junco-Quijandria[10] and
Ana Gonzales-Flores[10]

[1]Universidad Científica del Sur, Perú
[2]Universidad Continental, Perú
[3]Universidad Tecnológica del Perú, Perú
[4]Universidad Nacional Federico Villarreal, Perú
[5]Universidad Nacional Mayor de San Marcos, Perú
[6]Universidad de Ingeniería y Tecnología, Perú
[7]Escuela de Posgrado Newman, Perú
[8]Universidad Privada Peruano Alemana, Perú
[9]Universidad Cesar Vallejo, Perú
[10]Universidad Autónoma de Ica, Perú

Email: wauccahuasi@cientifica.edu.pe; olinares@continental.edu.pe;
krojas@utp.edu.pe; eflores@unfv.edu.pe; nbenites@unmsm.edu.pe;
aly.auccahuasi@utec.edu.pe; milnerdavid.liendo@epneuman.edu.pe;
julio.garcia@upal.edu.pe; gbernardo@ucv.edu.pe;
mcampos@autonomadeica.edu.pe; ajunco@autonomadeica.edu.pe;
agonzales@autonomadeica.edu.pe

Abstract

Currently, many upper limb prosthesis solutions are being presented, which are activated through the recording and processing of muscle activity, for which electromyography signal acquisition circuits are developed, where they are processed to find a level of activation necessary to activate various mechanisms that belong to the prosthesis. In the present work, we use a wearable device that performs the simultaneous recording of eight muscles, because it has integrated eight acquisition channels, and the device allows the recording and wireless sending of signals to various devices. As a result, we present an acquisition protocol where the registration of the arm muscles is performed, where we separate each of the signals that correspond to a particular muscle. The proposed method can be used in various applications where it is required to characterize the work of certain muscles in certain activities, which can create a database of the behavior of each muscle for certain activities, in order to improve the design of prostheses. As conclusion, we present how these signals can be used to recognize characteristic patterns using artificial intelligence techniques.

15.1 Introduction

We found works related to the recording of signals from the arm muscles, for which we resorted to the use of data acquisition circuits. The system uses solid state electrodes, which avoids the use of electrodes, having results similar to the records of the circuits that use surface electrodes [1]. Among the different circuit designs, we find some dedicated to the registration and control of acquisitions through wireless connection, which can control the devices. These are performed using techniques that offer us the IoT technology. This type of work is important because in emergency situations as suffered during the COVID-19 pandemic, many medical services are limited. Thanks to the implementation of these investigations, you can control equipment remotely as if a health personnel were manipulating it [2]. There are works where biomedical signal recordings are performed using wireless technologies, for which Wi-Fi networks are used as IoT, among others that make possible the elimination of wiring from the patient to the devices. With these methodologies, we can demonstrate the integrity of the data when using these wireless technologies [3].

There are works concerning the monitoring of patients by medical equipment with the use of invasive biomedical devices, which replaces major organs such as the heart, where iEEG signals are processed and recorded with the use of the Internet of Medical Things technology, together with remote systems [4].

The use of the Internet of Medical Things (IoMT) is being implemented in healthcare centers in order to diagnose certain pathologies, allowing interoperability between different devices for which IoMT is used, as well as a blockchain-enabled framework [5].

By applying the Internet of Medical Things applied in medical devices that collect information about the health of people through the Internet, where data are obtained from patients, in order to discover tumors by analyzing the characteristics through association rules, which uses a set of advanced features with which you can find brain tumors with respect to their degree, using techniques such as random forest and Naive Bayes, obtaining results that reveal that the set of features can replace traditional techniques with considerable accuracy and average [6].

We found works referring to the Internet of Medical Things (IoMT), which is considered as a mechanism for connecting medical equipment in order to share online medical data of patients for which parallelism techniques incorporating Hadoop, HDFS, and MapReduce strategies are used, which perform a parallel clustering mechanism with which to analyze gene expression data and evaluate the model in terms of computational complexity where the clusters constructed are used during parallelized classification using k-Se to confront with nearby algorithms, and be analyzed with standard classification algorithms, showing results of this experiment where the various parallel clustering algorithms are shown. We also show bit-level parallelism of the system with which it performs the parallelizations of convolutional neural networks in very large systems of graph processing systems where demonstrative results were obtained with the benefits and applications of IoMT applied to the real world [7].

We found works referring about accuracy and consistency about fetal monitoring interpretation, for which a framework for intelligent analysis and automatic interpretation of digital cardiotocographic signals recorded from fetal monitors based on Internet of Medical Things (IoMT) is developed, which implemented a method and system with which to evaluate fetal conditions during pregnancy showing accurate warning signs being used by media centers applied in patients from home with which the data segment is analyzed in a registry using automatic scoring functions such as Kreb's, Fischer, classification, where the results have been compared with the interpretations of obstetricians concluding that the results are accurate compared to traditional examinations [8].

We found works referring to the methods used to record and analyze the movements of the proximal muscles of infants, with the use of surface electromyography (sEMG), where 18 infants were used to record the trunk muscles of three-month-old infants, taking into consideration spontaneous movement and controlled postural changes. The total number of infants was divided into two groups according to motor performance for which a method was developed to eliminate dynamic cardiac artifacts in order to make an accurate estimation of individual muscle activation. At the same time, we took into account the quantitative characterization of the muscular networks obtaining the results after the atomized elimination of cardiac artifacts with which it has been possible to quantify the muscular activity of the trunk showing changes during the postural movements showing differences between high- and low-performance babies. In the muscle networks, constant change has been observed about the density of movements performed spontaneously between supine and prone position with a correlation of activities performed in individual pairs on the back muscles that are related to the motor performance of infants; so it is concluded that the sEMG analysis helps to detect differences between high- and low-performance infants [9].

We found works referring about the muscle movements of the arms of adults who have suffered some type of stroke and, as a consequence, have limited or no arm movements during the acute care process for which a prospective observational study is performed, for which 21 adults who have suffered a stroke during the previous five days of the level-1 trauma area are used, of which 13 are males and 8 females, of which among 11 patients, 7 males and 4 females had no observable arm muscle activity and in the remaining 10, 6 males and 4 females had detectable activity (MMT > 0), to which were placed dual mode sensors both electromyography and accelerometry in the anterior deltoid, biceps, triceps, wrist extensors, and wrist flexors of the injured arm, of which the number of muscle contractions, average duration time, amplitude, and co-contraction patterns of the injured arm were evaluated, and the number of muscle contractions, average duration time, amplitude, and co-contraction patterns of the injured arm were evaluated. The results obtained showed that muscle contractions were observed in five muscles of each patient located with electromyography (EMG) records, where contractions were identified after 30 minutes of monitoring of patients as MMT > 0 unlike patients with MMT = 0 who had a monitoring of 30 minutes for detection of five muscles during standard care, shown in the wrist extensors that have a greater amplitude in participants with MMT > 0 unlike MMT = 0. [10].

We found works referring to the motor pathways affected after stroke, which cannot be reliably detected about the motor recovery of the upper extremities for which transcranial magnetic stimulation (TMS) was added to be used during the clinical examination by using surface electromyography electrodes (sEMG) in TMS to obtain information about the muscle groups and corticospinal pathways; so the objective is to deter- mine the exact position for the placement of sEMG electrodes with which to record the activity of the flexor and extensor muscles where the optimization of sensitivity on the measurement of motor evoked potential (MEP) is used. MEP can be reduced or absent in patients who have suffered stroke, and at the same time, reference was made about the distinction between flexor and extensor groups; so for optimal flexibility multichannel sEMG was configured with 37 electrodes that surrounded the forearm where the determination of the optimal pairs demonstrates that potential amplitudes about the compound nerve action (CMAP) differentiating flexor and extensor muscles according to the distance of the electrodes responses are larger. So they measure the sensitivity about corticomuscular connections. Therefore, the responses about the muscle groups have been able to focus on the distance between the smaller electrodes; so it is concluded that with the application of this study, it has been possible to identify the best locations to place the electrodes during the clinical studies of TMS [11].

This work is dedicated to present a practical method to characterize the muscle activity by using the MYO wearable device that has the property of registering eight muscles simultaneously. The method proposed the recording mode and processing mechanisms in order to be used and exploited in various applications.

15.2 Materials and Methods

The materials and methods presented below are related to the description of the steps necessary to characterize the behavior of the muscles. One of the main components is the device that records the electromyography signals; in this sense, this component is important. The choice of the device ensures that the signal acquired is very useful and guarantees good results; In this sense, we have several options, from the development of hardware to the acquisition of commercial equipment, the proposed method uses the second option, working with the MYO model commercial equipment, which allows obtaining a clean signal.

Figure 15.1 shows the block diagram of the proposal, which begins with the description of the problem that is subject to analyze and characterize the behavior of the muscles of the upper limbs, and then we have the description of the method, which explains how we present the solution. Third, we have the protocol for using the method characterized mainly by the use of the MYO device. Fourth, we have the results of the method that explains the recording of the signals, and, finally, we present how we analyze the signals and how they can be exploited for different applications.

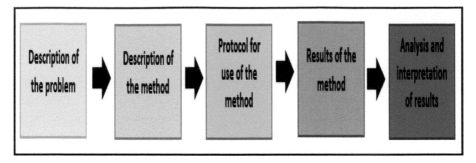

Figure 15.1 Description of the methodology.

15.2.1 Description of the problem

The problem we describe is related to the recovery of muscle activity. These problems occur in people who have suffered accidents and have their arm immobilized. As an alternative for recovery, we have myoelectric drive mechanisms. In both cases, it is necessary to know the state of the muscles to be able to present the patient with possible solutions to improve the mobility of the upper limb. The muscle, depending on the state it is in, has different behaviors, such as the level of activation of the signal as well as the duration of the activation. The monitoring of the muscles can be exploited when patients are in rehabilitation processes, where because of the lack of mobility of the muscle, they lost muscle mass, which is important to recover in its normal working state.

Starting with the description of the arm muscles, we find the brachioradialis or supinator longus, Figure 15.2, it is a long arm muscle in the external and superficial region of the forearm. It acts flexing the elbow and is also capable of pronators and supinators, depending on its position in the forearm. It is attached to the base of the styloid process of the radius and in the lower third of the lateral border of the humerus and in the lateral intermuscular septum.

15.2.2 Description of the method

The method presented is characterized by presenting the necessary steps to be able to take advantage of the device. For the benefit of patients, in order to better recognize the behavior of the muscle, we present below, how to record the electrical signals generated by the muscles, at the time of performing the work. For the purpose of having a clean record, it is necessary to know how the muscles of the arm are distributed, because the device to work is the

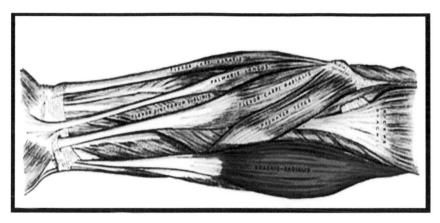

Figure 15.2 Brachioradialis muscles.

MYO, which has a bracelet shape, and in its content has incorporated eight solid state EMG sensors. This type of electrode is very important because it does not require electrodes to record the signal.

Since the device has a bracelet shape, it is important to describe how the arm muscles are formed, in order to know which muscles can be recorded; here is an example of how the arm muscles are distributed.

Figure 15.3, flexors: It is a muscle of the forearm that flexes the fingers. It inserts on the anterior aspect of the ulna and ends in four tendons in the third phalanx of the last four fingers. The four tendons are called perforators. Insertion: third phalanx of the last four fingers.

Figure 15.4, pronator teres roundus: It is a superficial muscle of the forearm, located in the external evening of the proximal and anterior region of the forearm; flattened and oblique.

The above figures indicate the first step for the application of the method; this stage is characterized by locating the muscles to be evaluated. This procedure must be performed every time the procedure needs to be started. Having located the muscle, the recording of the EMG signals is performed following the procedures.

The development of the architecture is a very important stage, it depends on the requirements of the problem to be able to choose the best hardware option that will be used as a mechanism for the acquisition of EMG signals.

Figure 15.5 shows the architecture of the method, where all the necessary elements are shown, from the analysis of the requirements to the final recording of the signals, including the configuration of the device.

These requirements are related to the needs of the patient, and these needs are related to muscle recovery to analyze possible mechanisms for

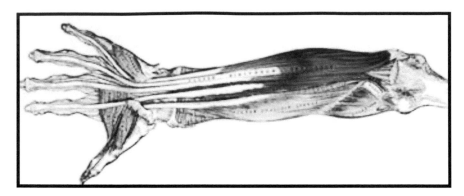

Figure 15.3   Flexor muscles.

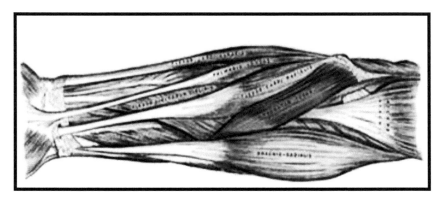

Figure 15.4   Pronator muscles.

future prostheses. This analysis of requirements allows to identify the hardware and software needed to meet the requirements.

Patients are also part of the methodology, because they are the ones who provide the problems and requirements. The analysis of the patient's situation is a fundamental task, both in the visual evaluation and in the physiological evaluation, where we can identify the patient's situation and possible solutions that can help to solve his or her health problem. In our particular case, which is addressed in this research, it is related to being able to evaluate the muscles that make up the arm, so that measures can be taken to benefit muscle recovery.

Having identified the patient and the requirements, the next task is to identify the best option for the hardware that can provide a solution; it is also necessary to identify the software needed to analyze the EMG signals. In our particular case, it is required to analyze the muscles located in the arm. In our case we resorted to the use of the EMG signal recording device, called MYO;

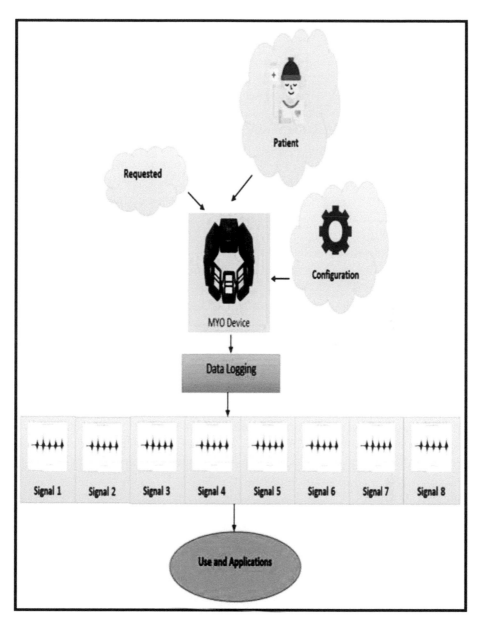

Figure 15.5 Method architecture.

this equipment consists of a set of eight solid-state EMG sensors, distributed in a circular shape, in a self-adjusting bracelet structure, which can be placed on the arm. This arrangement of the sensors allows a simultaneous recording of the eight sensors continuously. This device is very useful for its physical structure, because it does not require electrodes that are connected to the patient, but only places the bracelet, and, automatically, we are receiving the signals via wireless connection. It is important to note that the device must be configured. The configuration is dedicated to identify the device that can be a computer or a mobile device that is responsible for receiving the signals, as well as the MYO device's own settings that are the registration rate and transmission rate among other settings of the device.

After having configured the device, the next important procedure is also related to the type of processing, which can be online or offline. In our particular case, the device generates a file with the records made. The device delivers a file that can be with TXT or CSV extension. These files must be interpreted and analyzed so that they can be exploited. In our particular case, we worked with the computational tool MATLAB. Through this tool, we were able to interpret the file and recover the eight channels that correspond to each of the sensors that are related to the measurement of the muscle. The method proposes a mechanism for automatic reading and separation of the signals. Each of them is stored in a separate file, so that they can be exploited according to the requirements analyzed in the first part of the method.

15.2.3 Protocol for use of the method

The protocol for the use of the method is characterized by the necessary procedures to be able to use the MYO device, which ensures a clean acquisition of the signals. The most important element is the MYO bracelet. Being a commercial device of the THALMIC company, the device recognizes hand gestures through the acquisition of EMG signals by means of eight dry bipolar electrodes with a sampling frequency of 200 Hz. The MYO device contains an MK22FN1MO processor of the freescale kinetis cortex M4 model that executes tasks at a frequency of 120 MHz.

The characteristics of the MYO device are to be a usable and user-friendly device, capable of integrating with the different fields in which virtual reality is currently being developed, ranging from video games for training to the development of interfaces for sign language.

The EMG signal information and special data captured by the MYO device is transferred wirelessly via a Bluetooth low energy NFR51822 model. The MYO device sends packets containing a 128-bit universal unique identifier to the control service.

The MYO device is considered an interesting product, whose function is to allow the recording and visualization of muscle activity, which is very much in the style of the current system of muscle movement recognition.

Instead of relying on cameras or infrared sensors, the MYO is a simple bracelet that is placed on the forearm and through its sensors and detects the movement of the muscles in that area. In this way, we can take control of our muscles, and this device is also used for technological science, because this device also allows us to move objects by simply moving the hand, fingers, and arm from the distance we want.

The MYO has an inertial sensor of nine degrees of freedom, MPU9050 of which three GDLs correspond to the gyroscope, three axes to the accelerometer, and three magnetometers with which the position and orientation of the forearm are obtained with a sampling frequency of 50 Hz. It collects data of 20 milliseconds, which is more suitable to obtain the absolute position than the relative one. In addition to the aforementioned components, the MYO device has operational amplifiers for each electrode, for the preprocessing of EMG signals.

The methodology for the calibration of the MYO device is the customization of the device with a predetermined profile for each patient; this feature allows the acquisition of EMG data and its respective recognition of the hand gestures of any person. However, it is possible to improve this recognition through a calibration procedure in which a new customized profile is created with the user's EMG signal data. The calibration procedure consists of performing each gesture within the "Myo Connect" interface, which will automatically create the new profile once the calibration steps have been completed. The calibration interface is intuitive, indicating step by step the movements to be performed. Each time a new user puts on the MYO device, he/she must be trained with this "Myo Connect" interface in order to purchase and store a person's specific EMG signals in relation to his/her hand gestures. For proper functioning of the device and correct data acquisition, it is recommended that the main electrode of the Armband (in the MYO logo) be placed between the brachioradialis muscle and the extensor carpi radialis longus, i.e., parallel to the longitudinal axis of the humerus at the tip of the acromion, as shown in the following figure.

Figure 15.7 shows how the device is to be used for both calibration and operation; the logo identifier indicates the position of the main sensor. Figure 15.8 shows the configuration screen of the MYO device, with the parameters and a demonstration of the captured signals.

15.2.4 Results of the method

The results of the method are characterized by the signals obtained from the MYO device. To start using the MYO device as a device for interaction

Figure 15.6   MYO – EMG device.

with other elements, initialize either the custom profile (product of calibration) or the default profile of the device. The MYO is positioned on the
forearm and the wrist is extended to send a signal to initiate the synchronization of the bracelet. This is required each time the bracelet is placed
on the arm or removed, so that the device is calibrated in the positions of
the sensor depending on the location of the eight electrodes or connec-
tors on the muscles of the person. A few seconds should be waited while
the MYO stabilizes the temperature, in order to collect accurate readings
during acquisition. When the device is ready to be used, it will execute a
vibration and a "READY TO USE" message will be displayed on the computer screen.

For EMG data acquisition as well as position, it is done with the use
of MATLAB tool, using Mark Tomaszewski's Myo SDK MATLA BMEX
Wrapper package, which can transmit the MYO data at the device frequency,
200 Hz (EMG) and 50 Hz (IMU). This package after the construction of a
MyoMex-type target accumulates the data stream in the MyoData property,
as well as records properties of the gyroscope, accelerometer, and the EMG
signals of the eight bipolar electrodes of the MYO assumed in real time, having the possibility to access these data at any time and displaying the current
accumulated data.

In the following Figure 15.9, we can observe a scheme that corresponds
to the recording of the EMG signals in graphic form, when performing the
first gesture, in the previous image is shown for this test, the electrodes with

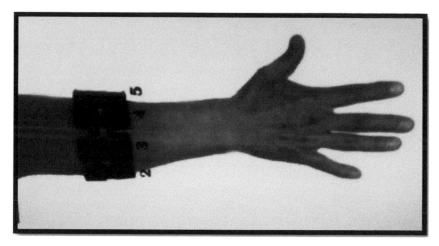

Figure 15.7 Configuration for calibration and use of the MYO device.

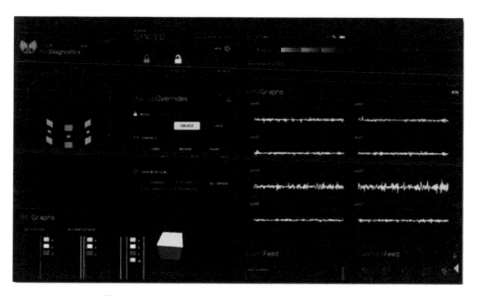

Figure 15.8 Main MYO device configuration screen.

the dominance of each detected gesture, where the red color indicates greater dominance, orange indicates medium level, yellow indicates low level, and blue indicates a null dominance.

In Figure 15.9, an example of the interpretation of the signals acquired by the MYO device is presented, where you can see the information of the

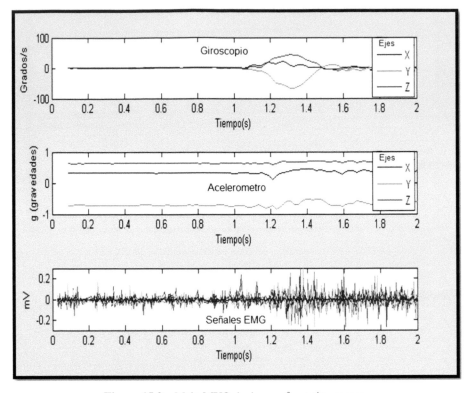

Figure 15.9   Main MYO device configuration screen.

different sensors such as the sensors that make up the gyroscope and accelerometer. In the upper parts, this information is important if you want to know the status of the signals when the arm is performing any particular movement, and in the lower graph, we find the visualization of the eight signals superimposed all together. In the analysis and interpretation of the signals present in the figure, it can be seen that when there is a variation of the gyroscope and the accelerometer, there is evidence of registers in the arm muscles, which indicates that an analysis can be made to characterize the different movements, having as complementary information the value of the sensors corresponding to the gyroscope and the accelerometer.

In Figure 15.10, the flowchart of the configuration procedure is presented. To be able to give conformity to the process of recording the signals, the processes to follow begin with the configuration of the device with the computer where the recording and storage of the files containing the values of the signals will be performed; mainly, the type of communication, the pairing of the devices, and the transmission speed are configured. Having configured

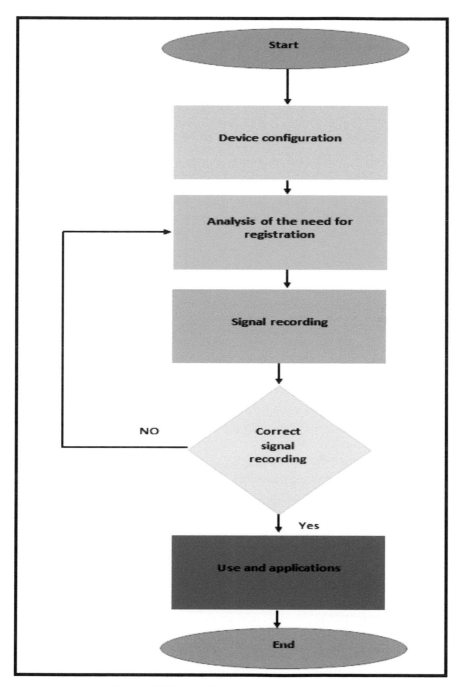

Figure 15.10 Device configuration flowchart.

the equipment, the next procedure is to check that the equipment can communicate and that a record is made, in order to verify that the generated file can verify the information of the signals. If it is verified that the file is in accordance and contains the stored values, it proceeds to make the record for the application to be used.

15.2.5 Analysis and interpretation of results

The last process of the methodology is characterized by the analysis and interpretation of the results of the process of recording and storing the signals, by means of the MYO device. In this stage, it is analyzed according to the initial requirements. This analysis is performed when it is necessary to verify the state of the muscle in order to be able to recover its capacities; it is known as an evaluation of the contexture of the muscle. When it is required to analyze the behavior of the muscle, when it is performing movements, the analysis is complemented with information from the gyroscope and the accelerometer, with which the behavior of the muscles is checked when it is in motion; this information can be used to perform simulations on the movement of the arm, after performing certain movements.

15.3 Results

The results we present are contextualized in three important aspects. The first from the perspective of the use of the MYO device, where it is possible to exploit its acquisition mechanism and the way it works, aided by the self-support of the device on the arm, added to wireless communication, provides independence at the time of recording, where the patient can move freely, so that the studies can be performed without connectivity problems.

From the perspective of the analysis of the captured signals, the results that we present are related to the exploitation and application of the signals. In the development of this work, two particular cases were presented, one when the intention is to evaluate the muscle to characterize the way it works and thus be able to achieve its recovery and a second application related to the characterization of the muscles, The recording of the signals can be used to model prostheses that work under the concept of myoelectric activation, where the intention is to characterize the behavior of the muscle when it performs certain movements. For this task, the device provides a lot of information by providing data that correspond to the gyroscope and accelerometer.

A third perspective is related to the various applications that we can use, as well as to describe certain characteristic movements. The device can detect by default up to five different postures, wrist flexion, wrist extension, fist, open palm, and gripper, as well as capture spatial data. With this information, it is possible to control other devices such as hand prostheses in very particular cases.

15.4 Conclusion

The conclusions we reached at the end of the research are as follows. We can indicate that in the study of myoelectrical signals, the muscles present certain characteristic movements depending on the action they are performing. From the movements of the arm itself when walking, to the recording of muscle activity when the arm is making certain movements, the activity of the muscles can be recorded so that they can be modeled later to create simulated movements that can describe the development of future prostheses.

Depending on the application and need to be covered, the device can be used. Analyzing the final arrangement of the sensors, the form of presentation in the form of bracelet, contributes a lot in the design of the support because it increases the applications to be self-supported and attached to the patient's arm, so that the patient can develop their activities in a normal way without the concern of being pending on the device. This particular feature of the device increases considerably when it is identified that the device transmits signals wirelessly to the connected devices. It is recommended to conduct a preliminary study of the patient's needs in order to make the best choice of hardware that can solve the problem.

Acknowlegdements

We thank the authors and co-authors for their participation and collaboration in the development of each of the processes, both in the experimentation and in the writing of this work.

References

[1] Auccahuasi, W., Rojas, G., Auccahuasi, A., Flores, E., Castro, P., Sernaque, F., ... & Moggiano, N. (2019, November). Analysis of a mechanism to evaluate upper limb muscle activity based on surface electromyography using the MYO-EMG device. In Proceedings of

the 5th International Conference on Communication and Information Processing (pp. 144–148).

[2] Auccahuasi, W., Bernardo, G., Bernardo, M., Fuentes, A., Sernaque, F., & Oré, E. (2021). Control and remote monitoring of muscle activity and stimulation in the rehabilitation process for muscle recovery. In Healthcare Paradigms in the Internet of Things Ecosystem (pp. 293–311). Academic Press

[3] Auccahuasi, W., Diaz, M., Sernaque, F., Flores, E., Aybar, J., & Oré, E. (2021). Low-cost system in the analysis of the recovery of mobility through inertial navigation techniques and virtual reality. In Healthcare Paradigms in the Internet of Things Ecosystem (pp. 271–292). Academic Press.

[4] Subash, T. D., Subha, T. D., Nazim, A., & Suresh, T. (2020). Enhancement of remote monitoring implantable system for diagnosing using IoMT. In Materials Today: Proceedings (Vol. 43, pp. 3549–3553). Elsevier Ltd. https://doi.org/10.1016/j.matpr.2020.09.816

[5] Li, X., Tao, B., Dai, H. N., Imran, M., Wan, D., & Li, D. (2021). Is blockchain for Internet of Medical Things a panacea for COVID-19 pandemic? Pervasive and Mobile Computing, 75. https://doi.org/10.1016/j.pmcj.2021.101434

[6] Khan, S. R., Sikandar, M., Almogren, A., Ud Din, I., Guerrieri, A., & Fortino, G. (2020). IoMT-based computational approach for detecting brain tumor. Future Generation Computer Systems, 109, 360–367. https://doi.org/10.1016/j.future.2020.03.054

[7] Sridhar Raj, S., & Madiajagan, M. (2021). Parallel machine learning and deep learning approaches for internet of medical things (IoMT). In Intelligent IoT Systems in Personalized Health Care (pp. 89–103). Elsevier. https://doi.org/10.1016/b978-0-12-821187-8.00004-6

[8] Lu, Y., Qi, Y., & Fu, X. (2019). A framework for intelligent analysis of digital cardiotocographic signals from IoMT-based foetal monitoring. Future Generation Computer Systems, 101, 1130–1141. https://doi.org/10.1016/j.future.2019.07.052

[9] Hautala, S., Tokariev, A., Roienko, O., Häyrinen, T., Ilen, E., Haataja, L., & Vanhatalo, S. (2021). Recording activity in proximal muscle networks with surface EMG in assessing infant motor development. Clinical Neurophysiology, 132(11), 2840–2850. https://doi.org/10.1016/j.clinph.2021.07.031

[10] Papazian, C., Baicoianu, N. A., Peters, K. M., Feldner, H. A., & Steele, K. M. (2021). Electromyography Recordings Detect Muscle Activity Before Observable Contractions in Acute Stroke Care. Archives of

Rehabilitation Research and Clinical Translation, 3(3), 100136. https:// doi.org/10.1016/j.arrct.2021.100136

[11] Munneke, M. A. M., Bakker, C. D., Goverde, E. A., Pasman, J. W., & Stegeman, D. F. (2018). On the electrode positioning for bipolar EMG recording of forearm extensor and flexor muscle activity after transcranial magnetic stimulation. Journal of Electromyography and Kinesiology, 40, 23–31. https://doi.org/10.1016/j.jelekin.2018.02.010

Index

About the Editors

Dr. Seema Rawat gained her Ph.D. in Computer Science in Engineering and currently she is working as Associate Professor in Amity University Tashkent, Uzbekistan. She has more than 16 year of experience in research, teaching and content writing. Her areas of interests include big data analytics, data mining, machine learning. etc. She has to her credit 12 Patents and has published more then 80+ research papers in international journals and conferences (Scopus Indexed). She is editor and reviewer for many books and conferences. She is First women to be part of the IP colloquium program of WIPO Switzerland Geneva and is an Active Member of WEC (Women Entrepreneurial Cell). She has been a panelist and speaker in many programs at corporate and university level. She has received the "Faculty Innovation Excellence Award 2019" on the occasion of the World Intellectual Property Day. The award was given by The Secretary of Department of Science & Technology (DST) Govt of India. She has also worked as Faculty Advisor of "AICSC (Amity IEEE Computer Society Chapter)" under AUSBI (Amity University Student Branch of IEEE). She is an Active member in IEEE, member of ACM, and member of IET(UK), SCIEI and other renowned technical societies.

Dr V. Ajantha Devi is working as a Research Head for AP3 Solutions, Chennai, Tamil Nadu, India. She received her Ph.D. from University of Madras in 2015, and has worked as Project Fellow under a UGC Major Research Project. She is a Senior Member of IEEE, and has been certified as "Microsoft Certified Application Developer" (MCAD) and "Microsoft Certified Technical Specialist" (MCTS) by Microsoft Corp. She has more than 40 papers in international journals and conference proceedings to her credit. She has written, co-authored, and edited a number of books in the field of computer science with international and national publishers like Elsevier, Springer, etc. She has been a member of the program committee/technical committee/chair/review board for a variety of international conferences. She has five Australian Patents and one Indian Patent to her credit in the areas of artificial intelligence, image processing and medical imaging. Her work in image processing, signal processing, pattern matching, and natural language processing is based on artificial intelligence, machine learning, and deep learning techniques. She has won many Best paper presentation awards as well as a few research-oriented international awards.

Dr. Praveen Kumar hold s a doctorate and master's in computer science and engineering. Currently he is working as Professor in Astana IT University, Kazakhstan. Previously he was the Professor and Head, Department of Information Technology and Engineering in Amity University Tashkent Uzbekistan. He has 17+ years' experience in teaching and research. Awarded as Best PhD thesis, Fellow Member of Indian Institute of Machine Learning recognized by Govt. of West Bengal, Best Teacher of Big Data Analytics, Corona worrier Award, Awarded for Excellence in Research-2023, Certified Microsoft Technology Associate (MTA).

He has to his credit 09 Patents, 04 Copyright and published more than 0130+ research papers in International Journals and Conferences (Scopus Indexed). He has organized and Session chaired in 62+ International and Evaluate 06 PhD thesis Reviewed as an External Examiner. He is guiding 04 PhD and 01- Ph.D. Awarded in area of Artificial Intelligence, Big Data Analytics and Data mining.

He has visited various countries for various purposes like Uzbekistan, Tokyo Japan, LONDON UK, Paris France, Da Nang, Vietnam Dubai, Russia and Kazakstan.

He is an Active Senior member of IEEE, LMIETE, ACM, IET (UK) and renowned technical societies and associated with Corporates as Technical Adviser of DeetyaSoft Pvt. Ltd. Noida, MyDigital360.